Writing in a Milieu of Utility: The Move to Technical Communication in American Engineering Programs 1850–1950

Teresa C. Kynell

The Ablex
Communication, Culture, & Information Series
Eileen Mahoney, The George Washington University,
Series Editor

Communication, Organization and Performance
by Tom Dixon, Queensland University of Technology

Unglued Empire: The Soviet Experience
with Communications Technologies
by Gladys D. Ganley, Foreword by Marshall I. Goldman

Writing in a Milieu of Utility: The Move to Technical
Communication in American Engineering Programs 1850–1950
by Teresa C. Kynell, Northern Michigan University

News Media and Foreign Relations: A Multifaceted Perspective
edited by Abbas Malek, Howard University

Forthcoming:

Diplomatic Discourse: International Conflict at the United Nations—
Addresses and Analysis
by Ray T. Donahue, Nagoya Gakuin University, and Michael H. Prosser,
Rochester Institute of Technology

Mass Media and Society
edited by Alan Wells, Temple University, and Ernest A. Hakanen, Drexel
University

Telecommunications Law, Regulation, and Policy
edited by William H. Read and Walter Sapronov,
Georgia Institute of Technology

World Broadcasting: A Comparative View
edited by Alan Wells, Temple University

Writing in a Milieu of Utility:
The Move to Technical Communication
in American Engineering Programs 1850–1950

Teresa C. Kynell

ABLEX PUBLISHING CORPORATION
NORWOOD, NEW JERSEY

A portion of Chapter 3 originally appeared as "English as an Engineering Tool: Samuel Chandler Earle and the Tufts Experiment" in Vol. 25 (1), 1995 issue of the *Journal of Technical Writing and Communication* and is reprinted with permission by Baywood Publishers

Printed in the United States of America

Kynell, Teresa C.
        Writing in a milieu of utility: the move to technical
communication in American engineering programs, 1850–1950 / by
Teresa C. Kynell.
                p. cm. — (Communication, culture, & information studies)
        Includes bibliographical references and index.
        ISBN 1-56750-264-4 (cloth). — ISBN 1-56750-265-2 (pbk.)
        1. Technical writing—Study and teaching (Higher)—United States—
History—19th century. 2. Technical writing—study and teaching
(Higher)—United States—History—20th century. 3. Engineering—
Study and teaching (Higher)—United States—History—19th century.
4. Engineering—Study and teaching (Higher)—United States —
History—20th century.   I. Title. II. Series: Series in
communication, culture, and information studies.
T11.R96   1996
620'.0071'173—dc20   96-18487
                                                                CIP

Ablex Publishing Corporation
355 Chestnut Street
Norwood, New Jersey 07648

To Kurt Von Syppli Kynell

Always

Contents

Acknowledgments

Studying the history of technical communication has been an enlightening process, from my first raid on the Michigan Technological University archives to endless requests for out-of-print books. At times I felt almost like a detective in search of elusive clues as I attempted to make some sense of engineering English and the shifts that led to technical communication. The discovery process involved countless hours with the Proceedings of the Society for the Promotion of Engineering Education, as well as specialized journals and even the popular press. This book, thus, is a synthesis of not only my own archival research, but of countless conversations, kind "leads" passed along by friends, and certainly the intervention of special few.

I cannot adequately acknowledge everyone who participated in the gathering of this material. Those who supported my efforts and helped in the culmination of this project certainly know who they are and know, I hope, the extent of my gratitude. The librarians at both Michigan Technological University and Northern Michigan University deserve thanks. Some individuals, though, deserve special recognition.

For their kind assistance, I express appreciation to Jack Jobst, Elizabeth Flynn, and Dennis Lynch. All three provided feedback and encouragement when I needed it most. I thank also Merrill Whitburn who provided me with not only copies of out-of-print books, but with leads and encouragement as well. Thanks, too, to Nell Ann Pickett who, with generous spirit, kindly provided a cold reading of the early manuscript. I must acknowledge Wendy Krieg Stone, a graduate student whose help during the copy-editing process was invaluable. I offer special thanks to Bruce Seely, who spent countless hours discussing with me the history of engineering education. His contribution to this project is inestimable.

Thanks are also due to those people who, just by virtue of their constant belief in me and the project, made life easier. Thanks go to my family members, of course, for their support, and to Fiona Gibbons and Donna Silta, two women who understand. I thank, also, the following colleagues: Anne Youngs, Ray Ventre, and Margaret Faler Sweany (all three always listened). I also wish to thank two mentors, Meredith Cary and Joseph Comprone, who helped shape the course of my career.

I would also like to thank Andrea Molitor at Ablex and Carol Davidson (formerly at Ablex). Both women made the process of completing, copy-editing, and printing this manuscript much easier.

Finally, I wish to acknowledge the one person without whose support I never would have finished this project. His encouragement, generosity of spirit, and love made all the difference. Thank you, Kurt.

Preface

The present study of the move to technical communication in American engineering programs is by no means an exhaustive treatment of the topic. The history of technical communication, as it evolved in American colleges and universities, is arguably difficult to sort out because of a variety of curricular influences, including engineering, science, business, and even medical and legal writing. What I have attempted is not an individual analysis of each of these very different disciplines which would require a much longer work. Instead, by focusing only on shifts in American engineering curricula, I hope to demonstrate how recognizable patterns in technical communication may have emerged. Piecing together our history is vital if we, educators, students, and scholars, are to recognize the unique features and evolution of this discipline—a discipline growing in popularity and stature. First, by evaluating one hundred years of engineering curricular movement, I hope to demonstrate how shifts in engineering English education provided a means for pedagogical experimentation resulting in a recognizable move to technical communication. Second, by isolating the hundred year period, 1850 to 1950, I hope to establish a foundation for future research into our evolution as a discipline. While several friends and colleagues urged me to take the study to "the present," I resisted because the shifts in technical communication *since* 1950 constitute a different study. If an analysis of the period 1850–1950 demonstrates *the move* to technical communication in American engineering programs, then analysis of the period 1950 to the present will likely demonstrate *the move* of technical communication into our English and Humanities programs as a distinct major. Such a study will surely set the stage for what is perhaps the more exciting period of our history—the future.

Introduction

Past research into the history of technical writing indicates, primarily, a preoccupation with Medieval and Renaissance styles and forms of writing, in addition to a steadily growing interest in some aspects of American contributions, including American scientific writers, authors of early technical writing textbooks, and specific 19th- and 20th-century industry-related cases. Although all these selections, both books and articles, collectively contribute to our as yet incomplete understanding of what constitutes the roots of *technical writing*, none have shown conclusively how technical writing as a discipline in America evolved and developed. In his recent bibliographic essay, "Studies in the History of Business and Technical Writing," William E. Rivers (1994) noted that "the value of historical studies has grown" and that recent works "trace patterns of influence and change over time" (p. 7). Although this is certainly true, as I demonstrate in a brief review of the literature available on technical writing's roots in American curricular patterns, few if any of those works have analyzed the *evolution* of technical writing in this country as it grew out of engineering English requirements from roughly the turn of the century to 1950. Indeed, the studies of technical writing's historical antecedents in America tend to rely on either the curricular patterns of one institution[1] or early technical writing textbooks and/or textbook authors.[2] Again, these materials are valuable and in fact inform this study to a degree. None of those works, however, focuses on or highlights the importance of engineering education as the central locus for the development of technical writing. More importantly, none of those works has brought together specific curricular patterns, texts, and writers, and analyzed them while tracing engineering educational patterns as they emerge in the proceedings of the Society for the Promotion of Engineering Education (SPEE). By examining the place of English in the curriculum of engineering students from the mid-19th cen-

---

[1] For example, see Grego's "Science, Late Nineteenth-Century Rhetoric, and the Beginnings of Technical Writing Instruction in America" (1987) an article that examines the emergence of technical writing at Penn State. See also Russell's (1993) "The Ethics of Teaching Ethics in Professional Communication: The Case of Engineering Publicity at MIT in the 1920s."

[2] See, for example, Schmelzer's (1977) "The First Textbook on Technical Writing" and Moran's (1993) "The Road Not Taken: Frank Aydelotte and the Thought Approach to Engineering Writing."

tury to 1950, as well as the influences of texts, their authors, and leading educators, I demonstrate how *technical writing* evolved and developed as a distinct English discipline in American engineering programs.

Thus, this introduction commences with a brief look at existing historical sources on technical writing, citing key articles and books that examine technical writing's emergence in America. A look at the SPEE, its role in establishing engineering curricular patterns, and its invaluable contribution to understanding how engineering English requirements evolved follows. Finally, this introduction considers the distinctions between technical and business writing, arguing ultimately that the discipline of technical writing is appropriately situated historically in an engineering context.

ESTABLISHING TECHNICAL WRITING'S HISTORY IN AMERICA: A REVIEW OF THE LITERATURE

Recent research into connections between technical writing and composition/rhetoric have proven useful,[3] but in order to understand more fully the emergence and development of technical writing, we must first consider the historical foundations of technical communication in this country. Three bibliographic pieces—Brockmann's (1993) "Bibliography of Articles on the History of Technical Writing," Moran's (1985) "History of Technical and Scientific Writing," and Rivers's (1994) "Studies in the History of Business and Technical Writing"—all provide selections from technical writing's past[4]; some of the articles are, in fact, worth considering.

First, Walter James Miller's (1985) "What Can the Technical Writer of the Past Teach the Technical Writer of Today?" in Cunningham and Estrin's *The Teaching of Technical Writing*, provides some sense of the evolution of technical writing by tracing examples of technical writing from the ancient Greeks to present times. This connection is potentially important in technical writing's evolution, particularly given the current interest in the humanistic aspects of technical discourse. Rhonda Carnell Grego's (1987) "Science, Late Nineteenth-Century Rhetoric, and the Beginnings of Technical Writing Instruction in America" looks at a much narrower subject—the growth of technical writing at Penn State University. This study offers useful insights into the contributions of both engineering and English faculty to that development,

---

[3] For example, see Allen's (1992) "Bridge over Troubled Waters? Connecting Research and Pedagogy in Composition and Business/Technical Communication" and Reynolds' (1992) "Classical Rhetoric and the Teaching of Technical Writing."

[4] Rivers notes a few other bibliographies that, because of their emphasis, do not inform this particular evolutionary study. See *Business and Technical Writing: An Annotated Bibliography of Books, 1880–1980* by Alred, Reep, and Limaye (1981). See also Hull's (1987) *Business and Technical Communication: A Bibliography 1975–1985* and see Zappen's (1987) article "Historical Studies in the Rhetoric of Science and Technology."

with Grego arguing that "late nineteenth-century rhetorical theory and composition practice evolved into early technical writing theory and texts" (p. 64). Her linkage of technical writing to composition pedagogy is especially relevant to this study. Russell Rutter's (1991) "History, Rhetoric and Humanism: Toward a More Comprehensive Definition of Technical Communication" examines essentially the technical communicator as "liberally educated generalist," tracing developments in the field of technical writing from the classical period to the present. This piece, although useful, covers a very large period of time and devotes only a few pages to shifts in technical writing in America from the 19th to the 20th century, a key period for studying the evolution of the discipline. Mary Rosner's (1983) "Style and Audience in Technical Writing: Advice from the Early Texts" on the other hand, looks at textbooks by Samuel Chandler Earle and Sada Harbarger, both key figures in the early evolution of technical writing in America. Rosner's brief analysis concludes that contemporary technical writing texts have not changed a great deal since the early texts.

Two recent books are also notable for their historical approach to writing pedagogy. First, David R. Russell's (1991) *Writing in the Academic Disciplines, 1870–1990: A Curricular History* accurately places key shifts in the early evolution of technical writing in the engineering departments of land grant and technical colleges. Although Russell devotes only select portions of his work to specific issues related to the development of technical writing, by placing technical writing in the context of general writing instruction in the United States, he is able to demonstrate the variety of curricular patterns that made up English instruction during the late 19th and 20th centuries. Second, Katherine H. Adams's (1993) *A History of Professional Writing Instruction in American Colleges: Years of Acceptance, Growth, and Doubt* traces the 20th-century partitioning of specialty topics in English departments across America. Although Adams does a fine job of chronicling the emergence of separate disciplines such as creative writing, journalism, and technical writing, her overview of the origins of professional and technical writing in American colleges relies heavily on anecdotal and archival materials derived from individual institutional settings. This, in addition to an analysis of early technical writing texts, is useful in piecing together the emergence of technical writing at select colleges, but does not reveal the broad shifts within engineering schools that ultimately influenced the direction of technical education in the United States.

Finally, any study like this must acknowledge Robert J. Connors's article "The Rise of Technical Writing Instruction in America." Published in 1982, the article analyzed technical writing in a way that no writer had attempted before—by tracing *trends* in the discipline through analysis not only of social, political, and historical movements and their influence on education, but also through examination of shifts in engineering curriculum and how those shifts resulted in an educational milieu conducive to a technical writing course. By depicting trends, Connors hinted at the evolution of the discipline, but because material from the late 19th to mid-20th century was covered in an article-length examination, a fully developed evolutionary study was virtually impossible. My study of technical writing as a discipline in America departs from past works because few have tracked specific curricular changes as I have. My purpose is to show that trends in engineering education provided the environment in which experiments in writing instruction took place—experiments that led to the development of a new kind of English course. The best way, I believe, to examine those evolutionary trends is to trace them from the inception of the Society for the Promotion of Engineering Education (SPEE) to the mid-20th century when technical writing's place as a discipline was developed and viable.

THE ROLE OF THE SPEE IN TRACING
TECHNICAL WRITING'S EVOLUTION

Although historical studies of professional writing have been undertaken, none have traced the *move* to technical writing as it unfolded in the proceedings of the Society for the Promotion of Engineering Education, an organization that, according to Reynolds and Seely (1993), had two purposes: "a continuing commitment to the improvement of instruction at the classroom level... [and] recognition from other professional societies and from governmental agencies as *the* spokesman for engineering education" (p. 136).

The SPEE, which first met in 1893, addressed all manner of engineering curricular questions, including the appropriate duration of undergraduate work, the course content of core requirements, and the role of the humanities in the curriculum. This last issue was no simple matter. From the inception of the society to the end of this study (1950), no clear consensus emerged as to what kinds of humanities courses, and how many, best prepared the engineer for life in society. Engineering education, though, is the locus for a study of this sort because changes in engineering curricula provided an environment for the gradual move to a technical writing discipline.

In fact, SPEE proceedings reflect the Society's dedication to classroom practice, including not only improvements in standard engineering pedagogy, but in more difficult issues as well—like the proper place of English instruction in the curriculum. In fact, members of the SPEE were leading thinkers in the field of engineering, with a great number of engineering faculty choosing membership in the organization. Because members were both teachers and researchers in the field, they brought original thinking and unique perspectives to engineering education. The proceedings of the SPEE serve as a surrogate for engineering education in America, a consistent record of the genuine interest and continuity of leading thinkers in the field.

Thus, SPEE proceedings provide an invaluable foundation for a study of this sort because, simply, the proceedings of no other group demonstrate such consistent attention to engineering curricular matters. In fact, Monte A. Calvert (1967), in his *The Mechanical Engineer in America, 1830–1910*, noted that the SPEE was a "powerful group... able in a few short years to gain complete control of curriculum, admission standards, and other basic constituents of engineering education" (p. 58).

The move to technical writing in America, in effect, unfolds in the pages of these proceedings, providing not only a context for some contemporary pedagogical practices, but providing as well a sense of the academic milieu in which technical writing emerged.

TECHNICAL AND BUSINESS WRITING:
ISOLATING THE DISTINCTIONS

Any study of technical writing's historical antecedents must also consider the distinctions between business and technical writing because the two are considered by some to be virtually interchangeable. Rivers (1994) clearly joins them as shown by the title of his bibliographic essay, and historical articles on both types of writing are combined in the bibliography that follows Rivers' article. This is not to suggest that the two writing disciplines have nothing in common. Indeed, they overlap in their attention to specific formats (e.g., letter styles) and their applications in the realms of business and industry, but similarities should not be extended too far beyond this. Business writing emerged in response to the specific needs of those involved in business-related enterprises and from the daily need for clear communication both inside and outside of corporations. As a result, the emphasis in much of business writing is on forms, letter-writing styles, and methods of improved communication. Technical writing, on the other hand, emerged in response to technology, most specifically in the need to communicate or describe that technology to an often less than sophisticated audience. Thus, technical writing is grounded in the mechanical or scientific arts and the need to explain the complexities of those arts as well as to produce user documentation so that technology, as it should, has relevance to and implications for the advancement of society. Business writing, on the other hand, is grounded in commercial enterprise, in the communication needs of organizations.

For example, Arthur B. Smith (1969), in his "Historical Development of Concern for Business English Instruction," noted that business writing instruction should involve (a) clear understanding of the actual wants in the minds of actual people; (b) knowledge of merchandise and their merits and demerits; and (c) [the] ability to use words so as to convey the merits and demerits of the goods to the customer (p. 41).

The obvious emphasis here on goods, merchandise, and customers clearly places business writing as a corollary to technical writing, but ultimately a very different undertaking than technical writing. The emphasis on report writing, technical description and definition, specifications writing, and user-manual writing clearly and correctly establishes technical writing's connection to technology. Although in a contemporary context the two types of writing share components, their evolutions as disciplines were quite different.

The purpose of this study, then, is to examine the pedagogical roots of technical writing in America by looking at the curriculum of engineering students from roughly 1850 to 1950. This crucial century appropriately frames my study for a variety of reasons. First, in 1850, significant shifts began to occur in the way engineering courses were designed in America. Engineering students, previously trained as engineering apprentices, faced changes in the way they would become professional engineers. After a variety of mid-century training and/or educational experiments, by 1880, most students became engineers largely after a four-year college education. Sweeping curricular change in a relatively short period of time brought with it status implications; engineers wanted to be perceived as the professional equals of doctors and lawyers. English instruction, interestingly, offered one means to answer the status question. Thus, engineering curricular experiments, beginning in 1850, led to English instructional shifts that influenced the move to technical writing in American engineering programs. Additionally, 1850 marked a period of great change in America. Technology was becoming a pervading fact of life, evidenced by armaments used during the Civil War from 1861–1865, as well as improving rail systems, the development of electricity, and the completion of the Brooklyn Bridge by the end of the 19th century.

By 1950, and with two World Wars completed, American society fully confronted the reality of technology, from DDT and sulfa drugs to radar and nuclear fission. Such technological growth, of course, meant great change for engineers, as well as the growing need to explain or describe those changes to an often less than knowledgeable audience. By 1950, in America, then, English in an engineering curriculum became the site of experimentation, with technical writing gradually evolving and developing to meet growing communication needs.

Framing this study from 1850–1950 acknowledges the importance of the period historically and establishes the critical importance of engineering curricular shifts and status-related issues that evolved as well during that period. Engineering, beginning in 1850, was in the process of *profession building*, as technical writing, arguably, has been since roughly 1950. Examining the move to technical writing in the milieu of changing engineering educational patterns will demonstrate not only how seemingly noncurricular issues such as status-related concerns made English instruction relevant, but how, in attempting to answer those concerns, a new discipline evolved.

Thus, understanding the gradual emergence of technical writing in America means understanding the academic environment in which the discipline was first taught—namely, English and/or engineering departments in predominately technological, four-year institutions. This study begins by looking at the roots of English instruction within an engineering curriculum, focusing especially on the initial importance of *culture* (literature) courses prior to the Civil War, to a move ultimately, by the turn of the century, to fewer literary courses and greater emphasis on writing pedagogy. Initiating this discussion, then, means examining the changes that occurred in engineering education in 1850—changes that ultimately produced an environment conducive to a greater emphasis on English in the curriculum.

To accomplish this, Chapter 1 provides background material from 1850–1900 on engineering education, analyzing how changes in an engineering curriculum impacted and influenced changes in English offerings at the same time, creating an evolving academic environment in which English courses and engineering goals were inextricably linked. Chapter 2 examines the need for a different kind of engineering English course, whereas Chapter 3 distinguishes between English as a *tradition* and English as a *tool*. Chapter 4 evaluates the role of cooperation in the move to technical writing, a role that led naturally to questions about the qualifications of teachers of this new discipline, a topic covered in Chapter 5. Finally, Chapter 6 looks at the emergence of a technical writing discipline in post-World War II America.

Chapter 1

1850–1900:
The Mission of an
Engineering Curriculum

ESTABLISHING AN ENGINEERING ANCESTRY

Analyzing the move to technical communication in American engineering programs means first addressing two foundational questions. First, why should technical writing, as we know it today, claim its ancestry in an engineering curriculum? Second, once the connection to engineering has been established, what can be said about the mission of both engineering colleges and departments in that process? Analysis of the curricular requirements for engineering students during the latter half of the 19th century reveals that engineering, like technical writing itself, struggled to discover what it should be and how it should fit in the greater body of academic institutions with distinctly more liberal culture concerns. In fact, the clash of technology and humanism, manifest by the mid-19th century, culminated in cultural and status concerns that not only preoccupied engineering educators for decades to come, but that established as well an environment in which technical writing ultimately played a role in engineering curriculum. In that process, technical writing was given a significant role by the engineers themselves.

This study examines the unique nature of technical writing, which emerged as a specialized discipline because of its foundation in a changing American culture—the explosion of technology and the rapid industrial advancement of the workplace. Certainly technical writing, as we know it today, overlaps both business writing and scientific writing, but technical writing largely emerged as a result of the engineer's increasing responsibility to communicate technological changes to a large and diverse audience—not just other engineers, but academics, lawyers, and the public. Thus, David R. Russell (1991), in his *Writing in the Academic Disciplines: A Curricular History* describes technical writing as "poised... between the pure sciences and industry [because] engineers 'translate' the result of research in the pure sciences into material products" (p. 120).

Perhaps what most establishes technical writing's foundation in an engineering curriculum is its late 19th- early 20th-century designation as *engineering English,* used consistently throughout papers included in SPEE proceedings, which was organized in 1893. Later, this study examines engineering English, its early roots in composition, and the role of English as a means to provide humanizing and socializing values.

As the following chapters point out, technical writing was born in uncertainty and self-doubt, often on the defensive in an academic environment that perceived the need for English as a humanizing more than a practical undertaking, due largely to lingering perceptions of vocationalism connected to an engineer's status vis-à-vis other professionals such as doctors or lawyers. Interestingly, the discipline of engineering would experience much the same self-doubt as it searched for its place in the academic marketplace, always acknowledging the utility of an engineering degree, but seeking also ways to humanize engineers, who were often perceived by other academics to have undergone little more than shop training.

Initiating any study of the move to technical writing in America, however, means first understanding the role of English in an engineering curriculum and the perceived role of a college education in this country during the last 50 years of the 19th century. What form did English courses take? How did engineering educators influence the content of English courses? How did concerns about the status of engineers create a need for English that ultimately allowed a course like technical writing to develop? These questions can only be answered by first examining significant shifts in engineering history from 1850–1890, because changes in the manner in which engineers were educated created an environment in which the value of English became apparent for a variety of reasons.

1850–1860: PATTERNS OF ENGINEERING EDUCATION

This discussion of the move to technical writing in American engineering programs commences in 1850, not because the first course appeared during that decade, nor because the SPEE, the foundation of this study, organized during that decade. Rather, 1850 is an appropriate place to begin this study because at about that time significant shifts began to occur in the way engineering courses were designed and offered in this country. Some brief background material demonstrates the relative importance of the mid-19th century to this analysis.

In the United States, engineering can first be traced to West Point, "the earliest college-level institution to offer engineering training" (T. Reynolds, 1992, p. 463). Rensselaer, founded in 1824 as "Rensselaer School," was the first private institution to offer engineering training in America (T. Reynolds, 1992). Aside from these two schools, prior to 1860, only a handful of departments or schools taught engineering in this country.[1] However, scarce academic offerings did not mean that few practicing

---

[1] T. Reynolds (1992) offers several figures: Charles Mann (1918) cited four schools or departments, William Wickenden (1929) noted six, Frederick Mavis (1952) found 7 or 8, Earl Cheit (1975) cited 11, and Lawrence Grayson (1977) believed there to be 12.

engineers existed. In fact, as T. Reynolds points out in his "Education of Engineers in America before the Morrill Act of 1862" (1992), engineers "were not at first trained in colleges. They learned their craft through apprenticeship, usually by finding a position with a practicing engineer and working with him in field and office for several years before seeking an independent position" (p. 460). In fact, even after 1850, most engineers still gained their experience through apprenticeship, clearly linking the field of engineering with vocationalism because the work had a greater connection to *shop trade,* or construction, in the field than it did to careers, such as medicine or the law, which were clearly connected to an academic setting. Second, during this same period, many classical colleges began to offer some coursework in engineering, but not fully established programs. Indeed, according to Reynolds, these *partial* or *select* engineering offerings were both popular and cheap because they mollified "critics by providing courses with some practical relevance without sullying the traditional bachelor of arts curriculum" (p. 468).

Another engineering curricular pattern that developed at this time was the addition of course work, primarily in math and science, to established Bachelor of Arts programs. "Recognizing," wrote T. Reynolds (1992), "that formal degrees carried more status than... 'partial' courses, many antebellum colleges pushed engineering instruction into their standard bachelor of arts curricula" (p. 470). In fact, the move to incorporate some engineering coursework into the Bachelor of Arts curriculum did not occur overnight, nor did it occur all over this country. Reynolds notes, in fact, that schools that incorporated engineering curriculum into a four-year Bachelor of Arts degree included a variety of southern schools[2] as well as some northern schools, including Columbia, New York University, Rutgers, and Rochester. Thus, from roughly 1850–1860, a young man seeking a career in engineering likely had either worked as an apprentice with a master engineer, had taken a few engineering or scientific courses, or had done both before pursuing work in the field.

However, changes were imminent for prospective engineers by 1860; the Land Grant Bill (Morrill Act) of July 2, 1862 would establish a permanent endowment of acreage and funding in order to "promote the liberal and practical education of the industrial classes" (Aldrich, 1894, p. 273).[3] During this period, then, a variety of patterns for achieving a career in engineering existed. As a result of the Morrill Act, land grant colleges (e.g Purdue)[4] developed in this country, while at the same time the educational patterns that preceded the Morrill Act (e.g. partial engineering programs and

---

[2] These schools included: the University of Maryland (1854), East Tennessee College (1840), the University of Mississippi (1848), the University of Missouri (1850–56), and the University of North Carolina (1853–54).

[3] A second national endowment—the Morrill Act of August 30, 1890—allowed for the "more complete endowment and support of the colleges for the benefit of agriculture and the mechanic arts" (Aldrich, 1894, p. 274).

[4] Other land grant colleges included state institutions such as the University of Michigan and the University of California, as well as universities such as Cornell. For a complete discussion of land grant schools, see Maurice Caullery's *Universities and Scientific Life in the United States* (1922).

apprenticeships) also remained. Furthermore, polytechnics evolved during this period (e.g. Rensselear). Reynolds notes at least six other polytechnics that had by this time been founded, including the Polytechnic College of Pennsylvania and the Brooklyn Polytechnic Institute. At other universities such as Harvard and Yale, science and engineering curricula were moved into separate schools.[5]

Two other institutes developed at approximately this time: Stevens Institute of Technology (1871) and Worcester Polytechnic Institute (1868). Emphasizing a very shop-oriented curriculum, these schools are worth considering because they offered prospective engineers yet another alternative in education. These schools promoted, more than a rigorous engineering curriculum, a hands-on approach to engineering reminiscent of the apprenticeship programs. Acknowledging the increasing acceptance of *school* as the place to become an engineer, these two institutes simply moved shop culture into a school setting. Notes Calvert (1967) in his study of mechanical engineering in America:

> Rather than being concerned with the education of professionals, these trade schools were devoted to training which would enable the student to immediately get a job as a foreman, superintendent, or even machinist. (p. 57)

The existence of these schools, in addition to the enduring classical schools, meant a variety of engineering educational patterns and schema in this country. An engineering education ran the gamut of very serious schools to a shop environment that clearly responded to the old apprentice argument that men needed a hands-on approach to mastering engineering principles.

However, by the time of the Morrill Act, engineering education was emerging as an alternative that could not be ignored. Shop culture schools such as Stevens and Worcester maintained the school ideal but often did not make room in the curriculum for liberal arts or culture courses, which were increasingly considered important components of engineering programs. In fact, Stevens and Worcester existed as reactions to the school argument, perceived by some as a compromise to an academic engineering plan—a plan that had as its impetus the Morrill Act. Many other engineering programs, though, did make room for culture—often in the form of English literature courses—because of lingering status concerns. Engineering, still perceived as a vigorous, physical undertaking, was considered by many as a vocational activity. Engineering educators, thus, increasingly turned their attention to avenues that would increase the status of engineers, in some cases adopting the patterns of traditional liberal arts institutions.

Traditionally, liberal arts universities catered to students who pursued higher education, not so much in pursuit of a profession, but in the desire to become *educated,* to become civilized and humanized. The Land Grant colleges, on the other hand, essen-

---

[5] Guralnick (1979) notes that faculty at Harvard's Lawrence School had titles such as "professor of *mining and metallurgy*," a title indicative of the utilitarianism that brought the school into existence (p. 123).

tially appealed to the middle-class desire to enter a professional trade that permitted a higher standard of living. Simply put, says Bledstein (1976) in his *Culture of Professionalism,* "by mid century, middle-class Americans were actively beginning to impose their expectations on the higher educational system" (p. 204). What this meant, too, was that the middle-class desire for a higher educational standard that met their needs was perceived by others in education to be a weakening of the status of an education in the first place. Henry P. Tappan, president of the University of Michigan in 1852, noted that the "commercial spirit of our country, and the many avenues to wealth which are opened before enterprise, create a distaste for study deeply inimical to education" (cited in Hofstadter & Smith, 1961, p. 491) These educational lines, so clearly drawn in the 19th century, point to a persistent pattern in our curricular past—namely, the desire to establish standards or norms for a curriculum to meet.

However, by 1865, some of the old patterns would begin to fade. Young engineers still held apprenticeships, but it was less common for those to be the sole means of engineering preparation.[6] In addition, T. Reynolds (1992) notes that the partial course pattern diminished in popularity because first

> an isolated course or two was unable to provide sufficient training for would-be engineers.... Second, the certificates of completion awarded by schools with "partial" or "select" courses... did not carry the prestige of a regular college degree. (p. 470)

By 1870, educators began to consider the importance—the status—that a four-year degree would bring to engineering education. In fact, the issue of status was a central factor in the mid-19th-century acceptance of engineering education in college settings.

1870–1880: STATUS CONCERNS AND TECHNOLOGICAL CHANGE

By 1870, status concerns were inextricably linked to the engineer's perception of himself and his discipline for a variety of reasons. Because engineers, prior to college course offerings in engineering, gained their expertise in apprenticeships and in shops, engineers were involved in, notes Guralnick (1979), "vocational preparation" (p. 120). The vigorous, physical nature of the work, often occurring outdoors, left the profession with a kind of hands-on, utilitarian association with decidedly negative connotations. Relative to *professional* degree programs such as medicine or the law, engineers, even as four-year programs in the discipline were beginning to evolve, were slow to lose the label of *vocational.* Thus, engineering educators by 1870 were very concerned with status—both for their students and for themselves. As Calvert (1967) notes, some fac-

---

[6] Calvert (1967) notes, in fact, that by the latter half of the 19th century, several firms, including General Electric and Westinghouse, "had developed regularized and systematic programs for giving technical graduates practical... experience without exposing them to the rough and tumble shop culture world" (p. 74).

ulty such as Professor Stillman W. Robinson of Ohio State University, were so concerned with the lingering connotations of vocationalism, that he dismissed the term *shop* for mechanical engineers, preferring instead, *mechanical laboratory*. "Why not," he asked, "call a chemical laboratory a medicine shop?" (p. 63). The comparison of an engineer's educational activities to a medical student's is obvious.

However, status concerns by 1870 were not just the manifestations of overactive engineering imaginations. Lack of recognition by other educators was real, to the extent that even some of this country's best-known four-year colleges and universities contributed to a marginilizing of engineering curriculum. In fact, notes Bruce Sinclair (1986) in his "Inventing a Genteel Tradition: MIT Crosses the River," the "Lawrence Scientific School at Harvard presents the most egregious example of... competition for cultural hegemony" (p. 4). The Lawrence School, founded and financed by textile industrialist Abbott Lawrence, was intended for "men of science applying their attainments to practical purposes" (p. 4). Almost immediately, continues Sinclair, the emphasis on engineering diminished, giving way to a "preserve of pure science ideology" (p. 4). Consider also, for example, the comments of Charles William Eliot, then President of Harvard, who wrote for the Atlantic Monthly in 1869 an article entitled "The New Education," in which he addressed not only the curriculum of both scientific and classical colleges, but also argued that the aims of both were quite different and did not necessarily mix:

> The [technical] student has a practical end constantly in view; he is training his faculties with the express object of making himself a better manufacturer, engineer, or teacher....This practical end should never be lost sight of by student or teacher in a polytechnic school, and it should very seldom be thought of or alluded to in a college... The practical spirit and the literary or scholastic spirit are both good, but they are incompatible. If commingled, they are both spoiled. (cited in Hofstadter & Smith, 1961, pp. 634–635)

At Yale, notes T. Reynolds (1992), engineering had its beginnings thanks largely to an endowment by industrialist Joseph Sheffield, whereas Dartmouth established the Chandler Scientific School, endowed by manufacturer Abiel Chandler. As in the case of Harvard's Lawrence School, both Yale and Dartmouth "were kept at arms's length from the main (that is, traditional liberal arts) college. They operated autonomously, and their faculty and students did not mix in classes or otherwise with regular college students" (p. 476).

In fact, universities, which continued to uphold an elitist notion of what it should offer students—the "rhetoric of liberal culture," says Berlin (1987)—largely set themselves up as antidotes to the potential dilution of the educational system the Morrill Acts permitted; suddenly a four-year degree was possible for a larger number of people. At roughly the middle of the 19th century, for example, John Henry Newman (1948) wrote *The Uses of Knowledge,* including one lecture entitled "Knowledge Viewed in Relation to Professional Skill."[7] Used widely in American universities,

---

[7] Newman's material was well known to college students. Houghton Mifflin's *Riverside Essays* (1913) contained an entire volume devoted to his writings on the role of a university.

Newman's lectures-turned-readings acquainted students with the purpose and goals of a university:

> This process of training, by which the intellect, instead of being formed or sacrificed to some particular or accidental purpose, some specific trade or profession, or study of science, is disciplined for its own sake, for the perception of its own proper object, and for its own highest culture, is called Liberal Education....They very naturally go on to ask what there is to show for the expense of a University; what is real worth in the market of the article called "a Liberal Education," on the supposition that it does not teach us definitely how to advance our manufacturers, or to improve our lands, or to better our civil economy; or again, if it does not at once make this man... an engineer;... or at least it it does not lead to discoveries... in science of every kind. (p. 54)

Perhaps unwittingly, Newman set up an either/or scenario for colleges and universities. That is, a university can meet its mission by providing a liberal education, or the university can train students for life in the workplace. In this academic milieu, mid-19th-century engineering faculty attempted to discover their place, often finding their virtual second-class status as educators unbearable.[8] "The solution to the problem was obvious: they must either increase the scientific content of their courses, in order to capitalize on the growing respectability of science, or increase their offerings in 'culture studies.' They did both" (Noble, 1977, p. 26). However, the increase in culture courses was gradual at best. Between 1870–80, those formulating engineering curriculum questioned the necessity of devising an educational framework for engineers that owed anything to the elitist liberal arts colleges. Although schooling was no longer a manifestation of engineering that could be easily dismissed, some in engineering education did not like the changes taking place. In fact, shop culture institutes such as Stevens and Worcester still existed as an answer to engineering as an *academic* undertaking. In the end, however, schools would supersede the shop approach, but why? The answer is tied not only to remaining status concerns, but to increasing technological demands as well.

Related to status issues were issues of increasing technological changes and the roles those changes would play in the education of an engineer. For example, by 1875, electricity was a reality, meaning, of course, the move to electrical engineering. Common sense, as some may have argued during the engineering apprentice days, did not provide an adequate foundation for working with electricity. Students might be taught how to *wire* an engine in a shop environment, but students required an academic environment if they were to learn how to *design* that engine. Thus, technological changes forced a reconsideration of engineering educational patterns in this country. In fact, by 1870–80, it became clear that scientific research was growing in popularity, to

---

[8] Veblen (1918), in his *The Higher Learning in America*, relegates engineering to the purely vocational, describing it as being "outside the academic field." He includes it in a list of other vocational undertakings such as "fitting schools, high schools, technological, manual and... other industrial pursuits, schools of 'domestic' science, 'domestic' economy and 'home economics' (in short, housekeeping)" (p. 191).

the extent, wrote Veysey (1965), "that most 'bright young men' were going into science" (p. 175). Engineers were quick to note the growing acceptance of scientific inquiry, and wrote Noble (1977), "from about 1870 on, the engineering curricula became distinctly more scientific and the focus of scientific study was shifted from laws of nature to principles of design" (p. 26).

It therefore followed that if an apprenticeship was not adequate for dealing with rapid technological change, then certainly partial course patterns or colleges that emphasized a shop-based curriculum were likely not adequate, either. Indeed, by 1875, whole new fields of technological endeavor were opening up, leaving the inadequately prepared engineer unable to compete. Thus, a fully realized college curriculum was both a means of dealing with changes in technology and a way of bringing to the engineering profession a desired modicum of status and respectability. In fact, noted Monte Calvert (1967), even simple words summed up the clash between shop training and education:

> The term "shop" was frequently replaced by professional educators with terms such as "laboratory" in descriptions of practical machine experience offered in the schools. Likewise, the use of the term "dynamical" instead of "mechanical" to describe the branch of engineering was favored by the educators probably because it avoided the root "mechanic" and gave mechanical engineering greater intellectual status. (p. 71)

1880–1890: ESTABLISHING AN ENGINEERING CURRICULUM

By 1880, engineering educators in this country were increasingly convinced of the need to move from *patterns* of education/training to *one* method of education/training—namely, a four-year degree. Once educators determined that a four-year Bachelor of Science degree was the proper means for engineers to initiate their career, obvious curricular questions arose. Furthermore, as engineers tried to decide what went into a four-year degree, some of the old questions remained. First, the Morrill Acts (the second Act coming in 1890), notes Connors (1982), "founded and promoted the land-grant agricultural and mechanical colleges [that would make a] college education available in the later 19th century to a hugely increased percentage of the population, colleges that were to broaden and specialize the college curriculum in many ways" (p. 330). This meant a great increase in the number of engineering schools,[9] as well as an inevitable split in what previously established classical colleges and the new engineering colleges believed their missions to be. Engineering colleges subsequently embarked on decades of self-doubt and self-recrimination, as administrators and faculty alike struggled with the ultimate goal of a graduated engineer. Was this student to be a person trained solely for professional responsibilities? Or was this person to be

---

[9] Noted David F. Noble (1977), "By 1880 there were eighty-five [engineering schools], and by 1917 there were 126 in the U.S." (p. 24).

both a professional and a cultured intellectual as well?

Second, if a university degree had become the accepted means for a person to become an engineer, how could the curriculum include the required mathematical, scientific, and humanistic curricula that educators clearly believed was necessary, as well as the hands-on work that ultimately prepared the student for the workplace? According to Reynolds and Seely (1993):

> Rapid technological change and restricted academic budgets were factors, but the main constraint was the limited time offered by four-year curricula. Educators increasingly set out to provide students with the scientific and mathematical principles of engineering activity and only limited hands-on experience. Employers would have to provide the experience that transformed students into practical engineers. This emerging consensus largely explains the appearance of the SPEE in 1893. (pp. 136–137)

In responding to the uncertainty about what constituted a proper engineering curriculum, the SPEE was organized and held its first meeting at the World's Engineering Congress in Chicago from July 31–August 5, 1893. This organization not only embraced the task of determining a common, or at least generally agreed on, course for engineering in America, but the proceedings of this organization for the next century provided insights into engineering curricular shifts. The period just prior to the first SPEE gathering was, wrote Connors (1982), "a rather dark time in the history of engineering education, a time when, by the schools' own later admissions, they turned out a large number of otherwise competent engineers who were near-illiterates" (p. 331). However, this did not mean that there was automatic agreement on writing-related issues. Although leading engineering educators were clearly concerned about their students' ability to write and communicate, the burdensome course load of most engineering programs led to a gradual decline in culture courses, particularly, literature and the belles lettres, because they were more easily trimmed away than required math and science courses. This, in turn, led to some obvious concerns about both the engineering students' cultural preparation and skill in writing, concerns that would preoccupy educators to the end of the 19th century.

In many ways English was not taught as part of engineers' training, but as a means to humanize them in the classical, liberal education tradition or to provide them with written communication skills. The key here is that in neither case—literature or writing—was English *tied* to the curriculum so that it had some purpose for existing. Thus, it is not surprising that such courses could in some cases be sacrificed to accommodate already overloaded engineering students. This is not to suggest, however, that all engineering educators embraced this view. Evidence exists, to the contrary, that key figures in engineering were very concerned with English in an engineering curriculum.

Mansfield Merriman, a civil engineering professor at Lehigh University,[10] conclud-

---

[10] Merriman (1893) noted in the same article that the Engineering Society of Lehigh University had published a quarterly journal since 1873, and that "the best training for a student in technical literary work is to prepare technical articles to be published over his own signature" (p. 264).

ed his 1893 essay in the first volume of SPEE proceedings, "Training of Students in Technical Literary Work," by noting that the "only way to learn to write is to write" (p. 264), predicting what became decades later the belief of many involved in teaching English. That an engineering educator as influential as Merriman made such a comment at the first gathering of the SPEE is worth noting. As a civil engineer, Merriman must have sensed the value of writing instruction in an engineering curriculum, his comment foreshadowing developments in engineering English instruction in the next century. Significantly, however, from 1893 to the turn of the century, not only did the Society for the Promotion of Engineering Education have no English faculty as members,[11] but concerns were voiced by engineers themselves. Moreover, the preoccupation of those writing about English instruction was whether students had time for or required greater familiarity with culture.

In 1897, SPEE President Henry T. Eddy noted in his opening address that culture studies were important to the engineer "because his preparation in the use of... writing... has been so meagre" (p. 13). He continued, however, that "instructors in the technical studies are apt to be impatient at the time and attention demanded by the culture studies as more or less of an obstacle and hindrance to what is rightly regarded as the student's main work" (p. 14). This statement accurately pinpoints the concern engineering faculty felt regarding the needs of their students. Students clearly needed both written and oral communication skills; yet, this knowledge of students' needs was certainly tied to English and thus in some ways bound to the concept of *culture*. In turn, because *culture* was associated with the established liberal arts colleges, engineers may have been willing and indeed impatient to abandon that aspect of education.

This was a rather ironic realization for engineering educators in general. On the one hand, engineering curriculum was deemed *vocational* because of the focus on utility and the lesser emphasis on the humanities. But on the other hand, engineers, attempting to find an identity in their discipline, wanted to develop a distinct curriculum, a literature, and a society that would remove them from the shadow of classical education, yet still provide them with the status they so desired. This curricular conundrum troubled engineers until after the turn of the century. Charles Carroll Brown, writing on engineering ethics[12] in 1896, addressed this "low estimation" of engineers—by the general population and by engineers themselves:

> The profession of engineering is too young, in this country at least, to have its status fully defined, and proper recognition of its high position in the world's work has been retarded by ignorance of its value and noble character. The writer is not one who considers engineering simply as a trade or handicraft, though most laymen and too many persons calling themselves engineers take that low view of its position. (p. 242)

---

[11] Connors (1982) noted in his article on the history of technical writing that an English faculty member did not join the Society for the Promotion of Engineering Education until 1905.

[12] Brown's call for a Code of Ethics for engineers is one of the earliest such calls. In addition to the practical aspects of such a code, he clearly believed that a code would also "legitimize" the profession of engineering in the eyes of both the public and engineers themselves.

ENGINEERING ENGLISH: COMPOSITION VS. LITERATURE

As engineering educators neared the end of the century, the issue of culture (and of the *English* role in culture) simply would not go away. In a lengthy paper highlighting the controversy as to whether culture studies classes had any relevance for engineering students, T. C. Mendenhall (1897), President of Worcester Polytechnic Institute, addressed the SPEE on "The Efficiency of Technical as Compared with Literary Training." After approximating the percentage of time students at classical colleges spent on literary subjects (67%) contrasted with the similar percentage of time (60%) engineers in technical colleges spent on scientific studies, Mendenhall concluded somewhat spuriously that "while no one can doubt the disciplinary effect of language studies, no one can claim that in the cultivation of the reasoning powers, the power of protracted, serious, and productive thought, they are at all comparable with mathematics and science" (p. 220) He concluded, therefore, that for students to spend well over half of their time on the "imperfect knowledge of languages" (p. 221) was certainly less valuable than scientific inquiry.

What is perhaps more striking than Mendenhall's attempt to quantify—and indeed qualify—the efficiency of science over language and/or literature was the 22 pages of "Discussion" that followed his argument. This commentary from faculty displayed a tendency to question the writer[13] and explore the positive influence of *culture* in the lives of engineers. Professor Johnson (1897), in a four page response to Mendenhall, called on the SPEE to move "toward narrowing the courses of study," because "the more liberally our students are educated, the more interest they have in society" (p. 231). By the 1898 gathering of the SPEE in Massachusetts, Robert H. Thurston[14] of Cornell reminded the society that the union of liberal and technical education "gave Sen. Morrill and his colleagues in the construction of the 'Land Grant Bill' stimulus, courage and persistence" (p. 104). Clearly the belief that engineers needed no cultural training to be professionals, a belief never universally accepted, was rapidly fading.

By the end of the 19th century, the means to provide *culture* in both the established and newer universities included history and philosophy and in particular, literature, most often at the expense of improvements in the teaching of composition. Although the teaching of philosophy was a popular means to offer students culture, English literature was, according to Veysey (1965), the primary means to this end. He noted, in fact, that "partisans of culture gradually installed themselves at a number of universities; by the nineties their voices echoed from inside many departments of English" (p. 183). As a

---

[13] Prof. Kidwell asked, for example, whether "Dr. Mendenhall is a classical scholar" (p. 225).

[14] Noted Calvert (1967) on the importance of Thurston's contribution to engineering: "Robert H. Thurston was an engineering educator… [who] was also accepted in the mechanical engineering elite. His ideas of what technical education should be had fully evolved by 1893, when he presented a nearly two-hundred-page paper to the ASME [American Society of Mechanical Engineers] on the subject of technical education in the United States. It was a complete system of education for mechanical engineering, and its presentation coincided with the formation of the Society for the Promotion of Engineering Education and the professional emergence of the engineering educator as a distinct type" (p. 104).

result, the growing emphasis on English as a cultural tool kept composition, particularly at the agricultural and engineering colleges, confined to a rigid series of drills and rhetorical exercises[15] that did little to increase a student's ability to write. English (literature), on the other hand, established itself as something more than a discipline—a means for change in the human psyche. Composition could not hope to compete.

The Harvard Report on the teaching of English, published near the end of the century, was a startling reminder of the inefficiency of English composition to turn out students who could write—and little wonder. While colleges and universities were concentrating efforts on *humanizing* students through criticism of, largely, English literature, composition remained limited to just a basic introduction to writing. In fact, noted Connors (1982), "freshmen composition requirements were almost universal, and the tacit assumption in engineering schools between 1880 and 1905 or so was that these first-year courses were all the introduction to writing that engineers needed" (p. 331). However, what sort of introduction were those students, nearly all college students during the period, getting? Unfortunately the composition class consisted primarily of formalistic rhetorical drills that were "an artificial motive for writing," said Gertrude Buck in a 1901 *Educational Review* article. To compound the problem, however, when the Harvard Report revealed "that the best students in the country attending the best university of its time had difficulties in writing" (cited in Berlin, 1987, p. 24), teachers such as Samuel Thurber (1893), again writing in the *Educational Review,* worried about the "inconsiderate adoption of useless books relating to composition and rhetoric, and the increased employment of formal methods that will impose enormous labor without practical results" (p. 383). In fact, composition remained formalistic for at least two more decades.

THE SPEE'S CALL FOR BETTER COMMUNICATION SKILLS

Ultimately, in the context of this study, the move to technical writing in American engineering programs can be located in the formalistic composition classes of the 19th century because this was often the only acquaintance engineering students had with writing. Before 1900, noted Connors (1982), there were no courses "that dealt with the needs of upperclassmen for knowledge of the writing demands of the engineering profession" (p. 331). An additional difficulty for engineering students was the perception of those guiding curricular change that engineering students needed composition to improve their writing, as well as literature to add culture to a course of study that was rigorous and rigidly scientific. The problem, of course, was that in a four-year program, already arguably too congested, engineering students did not have time to worry about writing and literature.

---

[15] Noted Connors (1986) in his "Textbooks and the Evolution of the Discipline": "Between 1860 and 1900 composition gave birth to that set of practical and theoretical doctrines that we now usually refer to as 'current-traditional rhetoric'" (p. 186).

Another complication in this equation, which certainly slowed the development of writing curricula in engineering schools, involved the implicit inability of educators to determine one standard curriculum for engineering students, a difficult task given the variations in high schools and the different levels of student preparation. In the 1894 SPEE proceedings, Frank Olin Marvin's[16] article, "Common Requirements for Admission to Engineering Courses," analyzed this problem by noting first that "many of our engineering schools have had their beginning as an offshoot from the old classical college" (p. 42), but also that the variety of coursework deemed necessary for an engineering student "makes us, as a body of teachers, liable to the charge of not knowing what we want" (pp. 42–43). Marvin's proposal for English requirements, including "considerable knowledge of the classics of English literature" (p. 49), included also the student's

> intimate acquaintance with the story, the plots, the characters and true meaning of at least ten masterpieces, chosen not only for their intrinsic worth, but also to illustrate the different periods of English Literature. The ability to write good English should be determined, not only from the formal English examination, but also from the papers presented in other subjects. The combination of a passing grade in English with execrable English in other papers is no rarity in the September of every year. (pp. 49–50)

Interestingly, Marvin pointed out the relative difficulties faculty experienced with their students' writing at the start of each new year, but he still suggested a curriculum that obviously emphasized immersion in literature. Although true that reading and writing about literature is one means to improve student writing, apparently faculty had not yet connected the importance of writing to the utility of the engineering profession. The irony is that many writers of this period did seem genuinely concerned that their students did not write well; many writers sensed the need for better communication skills. Even Marvin, later in the same article, argued for better writing skills, but noted that "the improvement desired does not consist of more knowledge of the rules of Grammar and the meaning of rhetorical terms… but rather implies an increased facility in the use of good English" (pp. 53–54).

However, the consistent emphasis on more literature and some composition did more than simply establish the different ways in which both forms of English were perceived (culture vs. utility). This arbitrary division also established a high (the classics) versus low (skill in writing) distinction, which engineering itself labored under as the established colleges were set up as institutions with lofty goals (ideas, abstractions, concepts), whereas the agricultural and engineering colleges were institutions with more mundane goals (an apprenticeship, a profession, a better standard of living). Writing of any kind—technical or otherwise—would have difficulty flourishing in an

---

[16] Marvin, Professor of Civil Engineering at the University of Kansas, in the same article, noted stiff admission requirements in foreign languages, with most of the Class A colleges he surveyed requiring from one to two languages; one school required three additional languages.

environment of occasionally arbitrary hierarchies established by educators who themselves were uncertain of where they fit in the academic world.

SPECIFICATIONS AND THE CALL TO UTILITY IN WRITING

By the end of the 19th century, several points became apparent. Technical writing as we know it today simply did not exist in American engineering programs, but *aspects* of technical writing did exist within composition classes. For example, some business writing (primarily business correspondence) was taught, but assignments that were uniquely technical—report and proposal writing, technical process and description writing, format devices, and audience considerations—were largely nonexistent. Why? The answer to that question appears to rest, ironically, in the dichotomous nature of English education up to this point. The English that engineering students were required to take, including both literature and composition, slowed the birth and evolution of technical writing, as educators argued among themselves that engineering students needed to become more *civilized,* as well as become more literate, so that they might communicate effectively on the job. Perhaps ironically, even today, technical writing has a similarly tentative existence in some departments, residing uncomfortably between the literary old guard and the new compositionists.

There were indications, however, that faculty began to perceive the value of a form of writing aimed at engineers and their needs. This was manifested in two papers delivered to the Society for the Promotion for Engineering Education in 1894 and 1895, papers that called for greater attention to the writing of specifications. At the New York gathering of the SPEE in 1894, J. B. Johnson, Professor of Civil Engineering at Washington University in St. Louis, made one of the earliest calls for teaching students to write specifications because "faults or mistakes in the specifications... often lead to litigation and delay" (p. 109). He went on to suggest a course that would include

> A description of the various papers, including the Advertisement, the Instructions to Bidders, Forms of Proposals, Plans, Specifications, Contract and Bond, which go to make up the set of writings which constitute the complete contract between the parties. (p. 111)

Johnson clearly perceived the need for the kind of writing left to, as he put it, "engineers of experience... and their legal advisors" (p. 109) because technology was rapidly increasing and American colleges were graduating more and more engineers. His description of a class in specifications sheds some early light on the kinds of concerns faculty had for their students, especially in their ability to communicate successfully on the job.

Then, in 1895, Ira O. Baker, Professor of Civil Engineering at the University of Illinois, wrote an article on specifications for textbooks that foreshadowed some of the elements of contemporary technical writing classes available today. He began by noting that his specifications were applicable to engineering reports, an early reference to the need for format in report writing. He followed with a lengthy numbered list of factors that bears a striking similarity to contemporary format considerations. In addition to calling for clarity in language, Baker argued persuasively for the arrangement of material in groupings, with "divisions and subdivisions... clearly indicated by distinctive headings" (p. 114). To that he added a list of format devices, including "sub-heads of a style of type different from that of the body of page... and subdivisions of articles by small capitals or italic side-heads" (p. 114). He went on to describe how supplements might be added to a text (or report), what type font was most readable, how "figures and diagrams should be inserted in the text" (p. 116), and how tables should be constructed and labeled. Clearly, Baker not only wanted to comment on textbook construction, but by presenting his recommendations to other faculty members, he presented an opportunity for later discussion of how this material might be relevant to engineering reports.

Although these two pieces by Johnson and Baker point to a gradual awareness of the necessity of technical writing and stylistic considerations, the 19th century ended without any real movement in higher education toward the development of a course that taught students these valuable and utilitarian communication skills. In many ways, for engineering English, the century ended very close to where it began—with educators trying to determine the form of English (literature or composition) most needed by engineering students. Although composition was taught, it usually constituted only the first year of the engineering curriculum, whereas literature, in an attempt to bring culture to the discipline, spanned three of the four years an engineer spent in college.

A review of engineering education from 1850–1900 reveals several trends that influenced English, and ultimately technical writing, curricula. The patterns of engineering education that existed throughout the decades, including everything from apprenticeships to partial university programs to land grant college programs, left engineering educators to sort out not only the mission of an engineering education, but the status of engineers as well. In fact, status issues, at the heart of the shift ultimately to a fully academic engineering environment, certainly contributed to the place of English instruction in an engineering curriculum because literature remained one, viable means for humanizing engineering students. Writing entered the conversation when engineers graduated with poor written communication skills. Thus, as educators attempted to establish a standard engineering curriculum, inevitable questions arose as to the place of both literature and composition in that crowded curriculum. By 1900, then, as influential engineering educators began to address the issue of written communication skills, calls for alternative English coursework signaled a shift to the right environment for technical writing to develop.

As engineering education moved into the 20th century, several factors changed. First, the deplorable writing skills of graduating engineers caused educators to reconsider the standard English curriculum. Second, composition itself began to experience curricular splits which influenced the growth of technical writing. Third, more specialized English textbooks began to address the unique needs of the engineering student.

Chapter 2

1900–1910: Establishing the Need for a Different Kind of Engineering English Course

As Chapter 1 demonstrated, engineering English curricular experiments between 1850–1900 simply were not working. Engineering colleges, as well as classically established colleges with engineering schools, were turning out "a large number of otherwise competent engineers who were near-illiterates" (Connors, 1982, p. 331). This, of course, was due to a variety of factors. English was really a course poorly perceived by engineering faculty and students alike. Because English instruction formed the foundation of liberal arts curricula, engineers, in an effort to construct a comparable foundation for engineering students, naturally included similar components. These components, largely composition and literature, were not popular.

In fact, the ongoing custom of requiring engineering students to take both literature (culture) and composition (communication) caused two distinct problems that potentially signaled the need for a different kind of engineering English course. First, students were not particularly interested in either literature or composition because they could not see the relevance of either to their future work. This was compounded, also, by their engineering instructors' similar belief that too much English distracted an engineer from his real work. Second, two separate English requirements—neither tied to engineering—made an already crowded curriculum even more congested. Engineering educators began eagerly searching for ways to make a four-year degree in engineering more compact and manageable. By the turn of the century in America, both of these factors would influence the way that composition for engineers was taught and perceived.

A survey of SPEE proceedings between 1900–1910 reveals three broad issues that contributed to a different kind of engineering English course: writing and the role of context, writing in a milieu of utility, and writing as a key to professional success. These issues, which emerged as engineering educators debated the best way to provide English instruction, mark the gradual emergence of technical writing as a reconceptualization of the composition course, a course that would better serve the needs of engineering students.

ENGLISH AND THE ENGINEER'S IDENTITY

Engineers, even at the turn of the century, were still often perceived as uncultured men in need of an identity due, in part, to the 19th-century apprenticeship pattern of engineering education. Neither liberal arts graduate nor vocational worker, engineers needed an academic focal point to establish them professionally and socially. Potentially, English offered a means to do that, but not in the form of literature or composition courses. As the SPEE proceedings reveal, engineers perceived the potential role of English in developing an engineering identity, although the realization emerges rather slowly. In fact, SPEE President, Ira O. Baker, in his 1900 opening address, not only noted the rapid growth of the engineering profession, but stressed his belief that engineers were still perceived as "relatively uncultivated" (p. 26). Baker may not have realized that his opening address would contribute to a decade-long dialogue on the identity of an engineer and, perhaps inadvertently, the role of humanities courses (notably English) in helping the engineer to forge that identity.

The behaviors, manners, and general level of cultivation of the graduating engineer remained a great concern of those establishing a standardized engineering curriculum. The old questions, first asked in the 19th century, remained. Should the engineer aspire to the virtually aristocratic behaviors of liberal arts graduates, or should the engineer embrace the role of craftsmen like those previously vocationally trained? The prevailing attitudes of the country at the turn of the century would not make it easy for educators to resolve this identity crisis.

On the one hand, curriculum, simply put, was inextricably linked to the *making of a man*. As such, engineering colleges wrestled with a curriculum that greatly reduced English requirements because English literature was tied to the humanistic. On the other hand, the general unpopularity of English made it difficult to justify for students who perceived their education as preparation for a career. Consider how, for example, perceptions of engineers could change in only 50 years. In the mid-19th century, when engineers were still largely studying literature as the *culture* portion of their programs, they graduated, says Noble (1977), with "the high status of gentlemen, and often had 'esquire' affixed to their name" (p. 35). Yet engineers would not enjoy this designation of aristocratic, educated gentlemen for long.

At the turn of the century, colleges took on clientele, "the broad range," wrote Bledstein (1976), "of people with middle-class aspirations" (p. 293). This acknowledgment of burgeoning democratic principles left the engineer vulnerable to criticism again. Suddenly, wrote Noble (1977), "the engineer...was scorned for being aristocratic" (p. 35). The political and social milieu was correct for appreciation of the skills of craftsmen, men who worked with their hands and their minds.[1] In fact, America was becoming an economic factor in global affairs. Concluded Baker in his 1900 opening address:

> We are now sending our manufactured products to all parts of the world, and if we are to have part in the commercial conquest of the earth, it will be because of the ability, the foresight, the wisdom of our own engineers. (p. 27)

Indeed, the desire to be educated for purposes of work or career enhancement was beginning to be perceived by some as uniquely American rather than "cosmopolitan and... closely allied with European civilization" (Veysey, 1965, p. 215). The debate strengthened over what a university should provide a student—a career or a means of thinking about the world. Writers alternately called for the practicality inherent in a *utilitarian* undergraduate degree, whereas others such as William Lyon Phelps called for a "community of intellectual interest. One does not dwell in a daily atmosphere of cloth and pork" (quoted in Veysey, 1965, p. 216). Engineering curriculum reformers, thus, had not yet considered English a means to synthesize these concerns. The time was right, however, Baker noted in his SPEE address, to begin looking ahead to the changes necessary in the next decade to secure the role, the identity, of the engineer—indeed, of the entire engineering establishment in America. As the decade evolved, English would play a pivotal role in helping the engineer to that end.

COMPOSITION AND THE ENGINEER

By 1901, after President Ira O. Baker had the year before enthusiastically asked SPEE members to look ahead to the future of the engineer, engineers instead were forced to consider the dismal communication skills of their graduating engineers. In fact, the problem was so serious that the SPEE, in 1901, formed a Committee on Entrance Requirements to address the minimum needs of engineering students in basic academic areas such as English composition and math. Freshman composition in nearly all engineering colleges remained virtually the only exposure students would have to writing. Although the committee stated that "in every case knowledge of [a] book will be regarded as less important than the ability to write good English," that same committee also recommended that "no candidate... be accepted in English whose work is notably defective in point of spelling, punctuation, idiom, or division into paragraphs" (p. 279). These recommendations reveal two important trends. First, educators nation-

---

[1] Indeed, even proponents of liberal education embraced the concept of the average working man, the embodiment of the physical and the ideal. Bliss Perry (1913), turn-of-the-century Harvard professor and prolific writer on the American mind, wrote that Americans were a "digging, hewing, building, breeding, bettering race... compounded of materials crude but potent" (p. ix).

wide were concerned that literature alone was not adequate for undergraduates; writing skills were becoming paramount. Second, the means to *correct* deficiencies in writing was still via drills in the mechanical features of writing. The notion that good writing is only grammatically correct writing spawned one of the most enduring and dominant forms of teaching English in the early decades of the 20th century in America—current-traditional rhetoric.

Current-traditional rhetoric, with its emphasis on correctness and task orientation, became a means for all educators to attempt to deal with the growing lack of basic writing skills in students. According to Berlin (1987), in his *Rhetoric and Reality,* current-traditional rhetoric found its original home at Harvard, the result certainly of the democratization of university admission standards. Many new students were ill prepared for the kind of writing they would be asked to do in college, so educators, reacting to a perceived wave of mediocrity, instructed their students in first "superficial correctness... and second, the forms [modes] of discourse" (p. 37). SPEE proceedings suggest that the same faith in mechanical correctness dominated engineering (composition) English as well, because engineering educators faced a twofold problem. First, students were graduating without good writing skills, in the most purely mechanical sense. Second, this lack of basic literacy fed the persistent notion that engineers were not cultivated, not as "civilized" as their liberal arts counterparts. Unfortunately, these combined perceptions only insured that current-traditional rhetoric would be a favored curricular choice in engineering colleges as well as some established liberal arts institutions.[2] Thus, the worse engineering students wrote, the more educators called for drills and mechanics, leaving engineering students, in particular, with a flawed introduction to such an important subject—writing. In fact, *writing*—the ability to construct a coherent and unified argument or report—seemed lost in the zeal of so many educators who elevated basic skills. Although mechanics are important, so are the intangibles of thinking and expressing thoughts cogently in writing.

If technical writing did break off from composition, reconceptualizing itself into a discipline that borrowed elements of compositional style, certainly the decade-long preoccupation with mechanical correctness as the primary objective of any writing course delayed that shift. With the focus of such a course so firmly located in the concept of *drilling*—forcing, if you will, the rules of English grammar into students' heads, thus insuring knowledge of rules, but not necessarily rules facilitating good thinking in written form—it is not surprising that both student and instructor disliked such a course. Indeed, as Gertrude Buck suggested in her 1901 *Educational Review* article, the missing factor in this strictly formula-driven approach to writing may have been the students' audience: "Even if thoroly [sic] indoctrinated with rhetorical formulae, the average student is conscious of no particular desire to produce for...

---

[2] The rhetoric of liberal culture, however, was taught more often than current-traditional rhetoric at selected East Coast colleges. The basis of liberal culture, as Berlin (1987) suggests, "was based on an epistemology that grew out of philosophical idealism... which held that all material reality had a spiritual foundation" (p. 44).

some unspecified and unimagined audience" (p. 376).

In fact, audience consideration may have been the missing factor in engineering composition courses, and may also have been a key contextual consideration for the eventual formulation of a technical writing course. Other factors also contributed to this different kind of engineering English course. The rapid growth in freshmen opting for degrees in engineering meant that the problem of relatively illiterate graduating engineers would continue to plague those developing a standard engineering curriculum; educators had to find a way to make the English class work for their students. Although clearly something needed to be done, many worried that the curriculum was "so loaded with required technical and scientific work" (Baker, 1900, p. 25) that students had no time for yet another English course that would add to what many believed to be too large an English commitment anyway. Clearly the time was right for a new kind of English course, a specifically *engineering* English course.

ENGLISH AND THE NEEDS OF ENGINEERING STUDENTS

Acknowledging the need for a new kind of English course meant first examining why the existing English curriculum was not working. In his opening address to the SPEE in 1901, President Frank O. Marvin did this by acknowledging the ongoing debate over whether an engineering degree was substantively comparable to a four-year, liberal arts degree; the inevitable comparisons between the two types of degree programs remained unrelenting. Marvin suggested that not only were the two types of degrees ultimately comparable, but in fact the engineering degree was more difficult, more demanding. Most important, his comment was a clear signal for engineers to stop comparing themselves to liberal arts students because, as Marvin said, "study for knowledge's sake may be stimulating to the few, but for the many there is needed the goal of a special calling to secure the close application that results in ability to concentrate one's energy to the attainment of a certain end" (p. 16).

Tied to this was Marvin's belief that the ability to excel as an engineer required a solid English background, involving more "than training in the writing of compositions, themes, forensics and reports" (p. 19). Marvin, and others guiding the course of engineering curriculum in America, saw benefits not only in severing any psychological ties that remained between technical and liberal arts universities, but also in rethinking some of the curricular choices made by engineering schools to duplicate liberal arts course offerings. Acknowledging, and highlighting, the difference between the two types of students meant also acknowledging that the kinds of courses which worked in classically oriented colleges might not work as well in technical institutions. Composition and literature might be courses serving the liberal arts well; however, were the courses serving the needs of engineers?

By 1903, T.J. Johnston, speaking before the SPEE in Niagara Falls, suggested that the current English curriculum was not serving students, as evidenced in the work of students and professionals alike. Although he advocated continued *drilling* of students, he also suggested that whatever schools were currently practicing simply was not working. In a surprise move, Johnston included in his paper not only a list of syntactical and grammatical errors made by students, but similar errors made by established engineers writing articles. Some of the errors Johnston noted are striking in that they are the same kinds of flaws contemporary technical writing instructors happen upon today:

- This operation is to be operated at every station.
- This discharge took place perfectly regular.
- The cost of superimposing a second floor on the present system would cost as much as the original cost of building the present system.
- The lapse of time which are here chiefly in question.

The list, much longer than the four examples included here, clearly upset Johnston, who lamented that "Wendell Phillips once said that there are only thirty seven jokes in all languages, but Mr. Phillips died before electrical engineers had written much" (p. 366).

Although Johnston's concerns did not go unheeded, SPEE proceedings reveal little discussion or activity on engineering English matters between roughly 1903 and 1907. This may have been due to several factors. First, Russell (1991) placed the first English faculty member in the SPEE at 1899, whereas Connors (1982), in his history of technical writing, believes that SPEE had "no members from English departments until after 1905" (p. 331). Regardless of the date, it seems clear that roughly from 1893 to 1910 the dialogue that occurred in the proceedings on the role of and teaching of English courses was virtually limited to engineering faculty members.[3]

THE ROLE OF ENGINEERING FACULTY IN
ENGLISH INSTRUCTION

This single factor—that educated and trained engineers for roughly 17 years determined the course of English instruction in engineering colleges across America— reveals a great deal about the slow evolution of technical writing as a distinct discipline, as well as much about the potential inadequacies of English education early in the 20th century. Simply put, the men trying to sort out the place of English in an engineering curriculum clearly acknowledged the necessity of teaching writing, but were, as has been previously noted, sometimes poor writers themselves. Also, even though the

---

[3] There exists little doubt that English, prior to the second decade of the 20th century, was taught by non-English faculty. A case in point was the Michigan College of Mines (now Michigan Technological University). The school, established in 1885, would list no English faculty in university catalogs until 1915, when Arthur D. DeFoe, with an A.B. from the University of Michigan, would be listed as an instructor in Technical Writing and Spanish.

necessity of writing was implicit in nearly all the SPEE comments, engineering faculty continued to wrestle with the issue of too many requirements and too much congestion in the curriculum. Clearly, their first concern was to turn out students who were highly skilled in their field; English had to be secondary in the minds of most. That is not to suggest that some engineering schools did not employ faculty members educated in English to teach writing. English curriculum, as Martin Telleen noted in his 1908 paper to the Society, was determined by the respective colleges.

Telleen not only suggested that a program loosely based on a writing-across-the-curriculum model had been proposed at some universities, but reinforced the notion that non-English faculty may have provided English instruction:

> Faculty cooperation, a plan which does away with a special English department and calls upon the teachers of all departments to enforce correct composition and expression... [knows] no dearth of advocates of the system. (p. 62)

In fact, non-English faculty contributed to early technical writing. James Souther (1989), in his "Teaching Technical Writing: A Retrospective Appraisal," noted that "in 1908, T.A. Rickard (himself an engineer) wrote what was likely the first book on technical writing,[4] *A Guide to Technical Writing*" (p. 4). This book was the product of Rickard's past work as editor of mining/engineering journals (mining and engineering his academic specialties). Connors (1982) observed that the book "dealt mostly with usage, meant more for practicing engineers than for college classes" (p. 332). A guide to style and language issues, Rickard's book had less to do with technical writing than it had to do with correctness. In many ways, the text was more a handbook, similar, in fact, to some of the business writing guides that also emerged at the time.[5]

Telleen's second major point was that English writing teachers at predominantly engineering colleges were pioneers. This is not surprising. Nearly all English faculty during the period from 1893–1910 were educated in the classics—literature—so that teaching writing was not considered a specialty. Indeed, many writers of composition's history (e.g. Berlin, 1987; Connors, 1982) have noted that composition was relegated to the category of *service* course and, as such, those trained to teach literature not only scorned a writing course, but were often held in low regard by fellow faculty members. To teach composition, then, in an engineering college, wrote Russell (1991), meant the teacher "held low status" (p. 122). But the contribution of those teachers of English to the formulation of a writing curriculum in engineering colleges played a great role in the reconceptualization of a composition course into a technical writing course, especially as they began to discern that students needed a reason to write.

---

[4] Because Rickard's book deals essentially with correction of mechanical errors, it might be more aptly described as a "technical editing" text.

[5] One of the earliest business writing series, a set of 12 small books entitled *Business English*, written by George Hotchkiss in 1909, were, according to Francis Weeks, virtually a correspondence course, emphasizing the "five C's" of business writing and other principles of what is largely the art of business letter writing.

CONTEXT AND THE MILIEU OF UTILITY

Applied mechanics professor, George Chatburn, may have indirectly elaborated on this need for a reason to write when he examined the role of context in a 1907 SPEE presentation, "A Combined Cultural and Technical Engineering Course." Chatburn opened with his belief that "technical engineering courses have in them the cultural value necessary and sufficient for any engineer" (p. 222). Perhaps inadvertently in his discussion of the point, he began to explore what was missing from engineering English courses, specifically composition courses, thus far. He argued for context in the form of cultural anecdotal material attached to lecture and suggested:

> It might be interesting to a carpenter to know that once the adze was a very useful tool, and to some of the early builders the most useful; it might be beneficial to know that the force of a tornado is not irresistible and that buildings can be made to withstand it. The first illustration is historical and therefore cultural, the second is technical but likewise cultural. (224–225)

How, Chatburn wondered, can engineers function in society when they cannot connect themselves historically or even sociologically to the people for whom they will work? They must know where they belong in the human scheme of things, and only cultural references will provide that. Engineers do not plan and develop their work in a vacuum; similarly, they cannot be expected to communicate in a vacuum either.

Chatburn, in arguing for cultural contextual material in engineering courses, supplied, perhaps indirectly, the reason students needed to write—a perspective for communication. Technical writing's gradual emergence may have been due to this relatively simple, but vital, recognition. Literature existed previously as a means to ennoble the engineer, but as a course it was not faring well in engineering colleges, largely because the students could not find the purpose in such a course. Engineering curricula were so oriented to the practical, the utilitarian, that most students undoubtedly found themselves searching for usefulness in all their courses. The steady stream of engineering graduates who could not write well points, also, to the ineffectiveness of composition courses for the same reason. Students tried to find the relevance to their own lives and future careers because they studied in a milieu of utility; it is not surprising that essays on topics not remotely connected to their engineering work had ceased to interest them.

Engineering students may not have required an English curriculum that catered to them, but they did need a reason to care about writing. Without a reason to write, writing does not exist as a communication tool. To write with a purpose, in context, means that any student who writes to another person must consider the interpretation of the recipient in order to ensure success. Writing to communicate also connects students to their peers, thus requiring those students to be ever mindful of the audience. This connection among people, which communication implicitly promotes, also provided what literature classes were somehow not providing in an engineering curriculum—cultural context. Thus, the gradual move to technical writing may have resulted, in part, from the decision to frame engineering English courses in a kind of professional, contextual model of the sort Chatburn described.

WRITING IN THE REAL WORLD

This notion of cultural context, as presented by Chatburn in 1907, was extended and reinforced by Martin Telleen in 1908. Telleen connected the relevance of context to the communication process when he wrote:

> The ability to express oneself correctly, clearly, and concisely has a pecuniary value, develops the power of initiative, of constructing, of composing—the real work of the engineer in all branches of his work—gives him the ability and facility of communicating his valuable experiences to his fellows. (p. 64)

In other words, Telleen looked at the mental, constructive aspect of engineering work in the communication process. What has been devised does not exist until produced; likewise, what has been produced does not exist for others until it is explained. When a reader comprehends successfully a piece of writing, communication has occurred, and a link has been formed between the encoder (writer) and decoder (reader). This process of encoding and decoding, sending and translating, establishes the humanistic stem so many developing a standardized engineering curriculum could not locate in literature studies.

In fact, Telleen (1908) perceived English's synthesizing power when he suggested that the "cultural, scientific, and technical [are served and] furthered by the English courses" (p. 65). Although Telleen had not, in 1908, presented a framework for a technical writing English class, he had begun to present some of the strongest arguments in favor of one. English previously had, perhaps, attempted to fulfill what engineering educators perceived as a cultural void in the lives of engineering students, but it had not succeeded in bringing elements of the scientific or technical to those students. Not only was a different kind of English course fully capable of providing this material in a relevant manner, but English embodied the means to advance engineers—in their chosen profession and in the perceptions of educators.

One way English could advance the engineer involved connecting English to technical studies within the context of real-world writing activities. Contemporary anecdotal material on the role of English in industry, according to Telleen, would not only provide a context for writing, but would also help the emerging engineer to see the relevance of communicating skills. Telleen described a technical report that one of his former students had to write while employed by the railroad. In relating this story, Telleen reinforced two points. First, English played a role in an engineer's professional life. Second, communicating on the job was often quite different from communicating at college. Telleen related the story of a student who was asked to write a report on railroad equipment, and who, after carefully drafting the document, was surprised that the

> Superintendent called him to task for presenting so cumbersome a report and then kindly picked out the essentials, which were put into a table....Both were compositions, but the purpose for which the report was intended commended the table of figures as being far superior to a composition in the generally accepted paragraph form. (p. 65–66)

The experience of Telleen's student provided some valuable insights into the need for technical writing, as well as the reasons why composition failed to work for so many engineering students. The utilitarian atmosphere of an engineering school did not implicitly promote either literature or composition. Even though the composition class prepared a student to draft a piece of writing that explained or described adequately, that class did not delve into such areas as format, tables (or any figures), and/or audience considerations. Thus, the composition class did not fail because of its focus; rather, the class failed due to the utilitarian environment in which it existed. This milieu of utility served as a constant reminder to students of the future—a future in the workplace.

A context for composition was not just the concern of technical colleges; the relevance of composition in any academic program was an issue taken up by educators and commentators nationwide as well. In 1906, Yale English professor Charles Sears Baldwin wrote that because composition really has no subject matter, it depends "on all subject-matters, [and] has for its essential function to serve all studies in general" (p. 488). The tone of this *Educational Review* article, occasionally reminiscent of Buck's (1901) piece, was one of hope for composition as an *open* discipline, which could be taught so that the student might bring "his studies into closer relation to himself" (p. 489). This belief that composition could serve the student led engineering English instructors like Telleen in 1908 to begin assigning technical topics in composition courses, thus signaling a period of change in engineering composition courses. In fact, the rudiments of a technical writing course began to evolve into what Telleen (1908) described as "technical composition" (p. 66).

The multiplicity of compositional styles, too, permitted the kind of experimentation that led to Telleen's technical composition. In fact, composition classrooms in nontechnical colleges were places of experimentation and some uncertainty as well. Berlin (1987) asserts that three types of compositional methods existed by 1910. Current-traditional rhetoric, with its emphasis on drills and correctness, continued to dominate, but the East Coast Liberal Culturists, "aristocratic and openly distrustful of democracy" (p. 45), were decrying freshman English instruction "even as they continued to provide it to students in freshman literature courses" (p. 46). A third, transactional model developed by F.N. Scott and his protegés at the University of Michigan, was "formulated as an alternative to current-traditional rhetoric" (p. 47). Not only was there no real consistency in composition education nationwide, but this "growth period," as Berlin describes it, meant that composition, flourishing in its development as an academic discipline, was constituted in such an open manner—its subject all subjects—that its lack of success at engineering colleges is hardly surprising. Composition in an engineering curriculum provided lessons in correctness, but it had to mean something more to students who ultimately cared about their ability to plan and design—not write.

This belief—that writing had to be tied to the future of an engineer—was also echoed by members of the popular media. A 1908 editorial in *The Nation* addressed the problems of composition in American colleges, while suggesting a means to bridge the gap that still existed between liberal arts and science. The unknown editorialist noted that instructors of literature, "the headspring of culture, and the teachers of science, which is proclaimed the only begetter of modern thought" ("English and Other Teaching," 1908, p. 253), should make writing relevant to the students' needs. Noting the problems that science students face, the writer suggested "Reports on laboratory experiments and other work in the sciences might be models of clear and orderly exposition; but they seldom are" (pp. 253–254). Clearly, when even those in nonteaching fields called for another approach to teaching English—particularly for those in technical fields who could not find relevance in studying composition or literature—the stage was set for change in engineering colleges.

This move to change English teaching was reflected not only in SPEE papers between 1900 and 1908, but also in the printed discussions that followed paper pre sentations in 1908. William Kent's (1908) "Results of an Experiment in Teaching Freshmen English" elicited a considerable response. The comments from discussion participants indicated a great desire for English to play a role in an engineer's professional life. The discussion notes, though, also indicated confusion about what a different kind of engineering English course should look like. Dean Benjamin,[6] for example, wanted to make sure students could write a good business letter, perhaps a facet of technical writing, but certainly not the primary focus of such a course. Professor Breitenbach indicated concern about "'highfalutin' nonsense about weird subjects on which they [students] never had any first-hand knowledge" (p. 90), whereas another professor was certain that "English training does not belong in the school of engineering" (p. 93). Martin Telleen (1908), though, concluding the comments, reminded the society that "the tendency toward practical work is evident," (p. 96) a clear suggestion that a new engineering English course was evolving and change was imminent.

WRITING AS AN EMPLOYMENT STRATEGY

One factor that influenced change was the role of future employers and what they considered to be important. Although graduated engineers were reporting to their instructors that writing was a facet of employment, few academics up until 1909 dealt with the issue of employers and their requirements. In 1909, however, Penn State engineering professor Hugo Diemer addressed the "Employers' Requirements of Technical

---

[6] SPEE discussion proceedings do not indicate at what institution participants teach and/or administrate.

Graduates" and seemed surprised himself to note that in the replies to his 720 mailed surveys (283 responded), 18 respondents (the second highest number) believed that English was an important "non-technical [study that] would best equip a candidate for our needs" (p. 174). In fact, respondents consistently reported the importance of good English skills, reinforcing Diemer's belief that "engineering courses, as now given did not tend to equip the student for ultimate executive or administrative work" (p. 177). This survey, the results of which were in keeping with prior calls for greater concentration in practical writing work, supported the changing engineering curriculum and the role of English in that curriculum in America.

In fact, members of the SPEE were not the only ones interested in exploring the connection between English and future employment. The lead article in the November 12, 1909 issue of *Science* magazine, A. T. Robinson's "The Teaching of English in a Scientific School," asserted that English "is intended to furnish a tool for business and professional life" (p. 658), once again reinforcing the role of future professional requirements. Clearly, just as engineering students needed a reason to write, a context for expression, so too did they need that ever-present component of an engineering education—professional success—related to their English studies. Engineers were oriented in their technical curriculum toward future professional success. Not connecting English to future employment isolated the topic, made it irrelevant to students, and removed from it all context. Literature had little or nothing to do with a professional future, and composition, although relevant in a general sense, also had little to do with the work of an engineer. However, as A.T. Robinson (1909) noted, success in English will come "with topics [the engineer] cares about and knows" (p. 659).

As the decade closed, engineers and faculty began to acknowledge two problems with engineering English education thus far. First, students, in writing either about literature or in the rhetorical modes, worked with topics that were unrelated to their future interests; thus, English lacked context for the kind of academic attention it required. Second, because English was not taught, analyzed, or approached as the other technical subjects were—with an eye to the future requirements of the employer as tantamount—English became a subject to endure, a curricular requirement isolated from the *real* undertaking, the study of engineering.

THE END OF A DECADE:
MERGING COMPOSITION AND TECHNICAL WRITING

As the first decade of the 20th century neared an end, English composition education remained a discussion topic not only in the engineering colleges, but in college in general. For example, letters to the editor of *The Nation*, a then-popular journal of current events, often dealt with the issue of poor writing skills for graduating college students. A March 3, 1910 letter from Columbia professor R.D. Miller proclaimed, "College English is bad, incredibly, intolerably bad" (p. 208). Engineers presenting papers and discussing ideas at the 1910 SPEE gathering in Madison, WI, evidently concurred.

John J. Clark (1910), Dean of the Faculty (International Correspondence Schools), presented a paper on "Clearness and Accuracy in Composition," decrying the lack of technical expertise and, more importantly, writing skill in textbook manuscripts sent to him. Whereas this, in itself, is not necessarily striking, Clark's suggestions for students and faculty as a result of his experiences are worth considering. He closed the decade of SPEE gatherings by arguing for a composition course that was more than drills and grammar. He favored a course that required the engineering student "to submit a composition relating to some engineering subject he was then studying or had just finished" (p. 340).

More specifically, Clark advocated a course in which students used English to demonstrate what they had learned in engineering classes. In order to explain how something is constructed or how a mechanism operates, a student would not only need to conceptualize the object or procedure, but "would reveal whether or not he really grasped the subject" (p. 341). In this way, engineering English could become a part of the curriculum that would offer students continuity and relevance, a course that would verify what those students had learned. English, rather than being a chore to survive, could become a means to check on the students' progress in all other classes.

In fact, the discussion that followed Clark's paper reinforced this notion that English could serve as a means of ascertaining students' proficiency in engineering courses. Dean Crouch, for example, recommended that students "describe some simple machine, [because] if they can describe this so that the professor of English can understand the machine, they can use good English" (p. 345). Later, Professor Walter H. Drane echoed his point by suggesting that students at the University of Mississippi submit weekly written reports that "are not simply technical in nature, but are descriptive" (p. 351). A third professor, H.L. Seaver, advocated taking "a Dover eggbeater and require an exposition of its easily perceived but almost indescribable movements" (p. 355). These calls for what would certainly today be described as technical description seem to be a natural juncture in the early move of technical writing away from composition.

Remember that in the late 19th century one of the closest suggestions of technical writing was one instructor's call for students to write specifications—a kind of technical description of a product and/or mechanism. By the end of the first decade of the 20th century, when the need for and move toward technical writing was becoming more evident, a kind of general description of products or mechanisms appeared to be a way, again, of verifying the engineering students' ability to digest and represent information to a specific audience. Students, by doing this, utilized one of the modes—description—they had learned in composition, but with a different application. This use of descriptive style applied to a technical subject (e.g., a mechanism) indicates how technical writing may have grown out of a composition tradition. Indeed, engineering faculty, in particular, may have reconceptualized the traditional engineering writing assignment so that what remained—elements of description applied to a technical subject—was a composition hybrid. Such a descriptive assignment still asked students to write, but it asked more importantly that they write on a topic that had immediate relevance to their futures.

By 1910, those guiding the future of engineering English in America seemed aware of several pedagogical factors that would ultimately prove to be foundational for a new kind of engineering English course. First, writing had to be tied to the students' interests if English was to play any meaningful role in their lives. Second, the way to tie English to engineering was somehow through engineering topics themselves, providing, then, a kind of real-world context for writing. And third, English writing should have some basis in the composition modes, but should ultimately have less to do with literature and more to do with engineering practices. What those thinking and writing about the role of English in an engineering curriculum may not have known was the number of changes in English offerings occurring at Tufts College under the guidance of Samuel C. Earle. As members gathered for the second decade of meetings of the Society for the Promotion of Engineering Education, few could have foreseen not only how English would become increasingly a preoccupation of the SPEE, but, more importantly, how "technical writing," as we know it today, would be presented as an alternative for all engineering students.

Chapter 3

1911–1920:
Distinguishing Between English as a Tradition and English as a Tool

If the first decade of the 20th century ended with a call for a different kind of English course, the second decade, 1911–1920, began and ended with two key figures who presented possible models for a technical writing course and the faculty who would teach such a course. In 1911, Samuel Chandler Earle gave his groundbreaking presentation on technical writing to the SPEE; in 1920, Sada Harbarger offered the SPEE the qualifications necessary for an engineering English teacher. These two figures framed a decade of questions for those teaching English in the engineering schools—questions about English requirements, English faculty, and, most importantly, the reasons for offering English in an engineering curriculum in the first place. Of all the questions engineering educators tackled, the following question highlighted the decade: Should English offerings reflect some of the traditions of the past, or should English be regarded as an engineering tool?

Of crucial importance to this decade was the SPEE's formation of the English Committee in 1914, a committee that wrestled not only with the need for English in an engineering student's life, but with the existing attitude toward English in engineering schools. The committee's findings, although not universally negative, were less than encouraging. Still, the committee's activities provided the impetus for some of the larger issues related to engineering English in this decade, issues that guided not only the direction of engineering educators, but also this analysis as well. Chapter 3 traces those issues, specifically non-English faculty and the emergence of more engineering English textbooks.

SAMUEL CHANDLER EARLE AND THE TUFTS EXPERIMENT

Earle resoundingly believed that English was an engineering tool; so few papers published before or since his 1911 "English in the Engineering School at Tufts College" have generated such lengthy discussion by SPEE members and engineering educators in general. Earle, a PhD in Philology and professor of English at Tufts, began teaching a rudimentary course in technical writing because, noted Russell (1991), there was a lack of "work teaching literature," not to mention the fact that Earle "liked the job and began to realize its [technical writing's] possibilities" (p. 122). Indeed, according to Connors (1982), Earle's role as a key figure in technical writing cannot be overstated.[1] In fact, Connors notes that since 1904, Earle had been teaching classes at Tufts "that were perhaps the first recognizable technical writing courses" (p. 332). His method for teaching engineering English clearly separated him from earlier writers on English education. Not only was his experiment very similar to the kind of technical writing taught today, but Earle's innovations also provide insight into the ultimate direction of the course in engineering colleges. Although others may have formulated a similar curriculum, indeed even teaching a course like the one Earle outlined, scant evidence exists that anyone constructed a foundational approach to English for engineers as did Earle. That his methods were clearly written and accessible, not to mention widely disseminated through the SPEE, lends credence to Connors's contention that Earle "was a true educational ground breaker" (p. 333).

In the first paragraph of Earle's paper, he referred to his English experiment as *radical,* suggesting that not only was he unaware of any other English faculty approaching the subject in a similar manner, but also that what he was proposing constituted a departure from the other attempts at modifying either composition or literature offerings to accommodate the engineering student. And make no mistake about it, Earle was not only an English scholar and a regular member of the SPEE, but a man truly interested in the question underlying English instruction at technical colleges: Why offer the subject at all when, in its present form, it functioned only to annoy student and faculty alike? In attempting to answer that question, Earle also presented an alternative that he knew would not meet with universal enthusiasm. Rather than continuing to use an English course as a means to culture, Earle believed English should become a technical tool—a tool accepted and agreed upon by engineering and English faculty alike.

In fact, engineering and English faculty by 1911 still had not come to a consensus on the role of English. Connors (1982), in his "Rise of Technical Writing Instruction in America," noted that Earle "condemned the attitude of English teachers that saw engineers as Philistines, to be proselytized to about the superior virtues of culture and literature over engineering" (p. 333). Although Connors may be correct in some instances, considerable evidence exists that English and engineering faculty wanted a

---

[1] Connors (1982), in his "The Rise of Technical Writing Instruction in America," for example, calls Earle the "Father of Technical Writing instruction" (p. 332).

way out of this conundrum. Engineering faculty wanted academic acceptance for their students, and English faculty, nearly all of whom were trained in literature, saw English literature as a means of providing that which was perceived as missing in the engineering student's curriculum, namely culture. Earle addressed this apparent impasse clearly and directly. English, he wrote, is perceived

> as the last bit of salvage from the arts course remaining in the engineering school and as the only means of true culture in a curriculum otherwise hopelessly practical. (p. 34)

English, he implied, remained in the minds of many the last vestige of the arts, the final link in all degree programs, the reason engineers could find a place among the liberal arts graduates. But Earle proposed a course "as broad, and as varied as that given students in arts," because "true culture comes not from turning aside to other interests as higher, but from so conceiving their special work that it will be worthy of a life's devotion" (p. 35).

Earle very clearly believed that culture was implicit in the engineering discipline. If not, logic would dictate that students must look elsewhere—outside the discipline—to find enlightenment or culture. To look elsewhere presupposes that disciplines are constituted not according to what they are, but instead according to what they are not. Thus, as long as engineering faculty and curriculum reformers sought ways to fill self-perceived gaps in an engineering curriculum, students could never hope to maintain a position of equal standing with liberal arts graduates because they would have been brought up through the academic ranks by people who were always implicitly attempting to justify their purpose in an academic environment.

This attitude, of course, is ruinous to student and teacher alike because some element of legitimacy is always missing. Earle argued that when engineering and English faculty alike recognized the rewards of English as a tool for the engineering student rather than a time-honored tradition, English would become relevant and enculturating simultaneously.

Earle proposed, quite simply, a systematic approach to the "constructive creation" of written material, emphasizing writing on technical subjects. Certainly, work approximating this had occurred already in engineering English or composition classes, but Earle stressed that it was a mistake to believe that a writing course in composition was adequate: "No lawyer, minister, dramatist, novelist, or poet would assume that because he had had general training in composition he could become a master of his peculiar form without special study" (p. 37).

Earle's comment about composition raises several points worth considering. First, prior to Earle, most writers who analyzed the problems of English in an engineering curriculum argued that overhauling the composition requirement might solve the problem. Earle was one of the first writers on the topic of English for engineers who suggested that the problem was not reconceiving the composition classroom, but reconsidering the composition requirement as the sole means to good expression for an engineer. Second, by placing engineering in a list that includes lawyers and professional writers, Earle presupposed the rank he believed due the engineer professionally. Professionals such as doctors and lawyers had always maintained a form of discourse unique to their career choices; was the engineer any different? Third, Earle provided a foundation for the engineering college's potential split with composition as the only means to writing available to engineering students by pointing out that the aims of composition and technical writing were fundamentally different.

THE IMPORTANCE OF OBSERVATION AND DESCRIPTION

Although Earle referred variously to his course as *technical exposition, engineering writing,* and *engineering English,*[2] its differences from the traditional composition course were clear. The course Earle envisioned completed the job of other English courses, "the final bringing of the training in composition into relation with the special problems of the profession" (p. 37). More than repetitious drills in correct grammar, mechanics, and spelling (areas that remained a preoccupation for many),[3] Earle proposed four separate abilities that would make English more relevant to engineers:

1. The ability to put into words an abstract thought.
2. The ability to describe, in writing, an object not present.
3. The ability to write for different audiences.
4. The ability to give a concept full treatment by demonstrating understanding in writing. (pp. 38–39)

The first two of these opportunities point to what may have been the virtual hybridizing of aspects of composition into this vision Earle had for a technical writing class. Borrowing on the rhetorical modes,[4] specifically description, Earle believed that the ability to describe in writing a concept, product, or object was a means not only of providing students with a workplace context for their work, but was also a way of informing the instructor—the engineering instructor—of the full grasp students either did or did not have of their topic.

In a composition environment, students were encouraged to describe objects, places, and people in an effort to teach them the value of detail. Taking that same mode of writing and placing it in the context of engineering was a natural extension of the work some students may already have had, thus bringing the material full cycle as

---

[2] *Engineering English,* throughout this study, refers to the English requirement for engineers—almost always a composition course.

[3] Spelling remained a major preoccupation of engineering educators, perhaps because so much documentation was handwritten, and poor spelling made the reader's job potentially more difficult. A 1908 SPEE paper by William Kent, a Syracuse University Dean, entitled "Results of an Experiment in Teaching Freshmen English," included a list of 204 misspellings he had collected from engineering students. He included "combusted" (for burned) because "the spelling may be right, but the word is a barbarism" (p. 81).

[4] Earle published, in 1911, *The Theory and Practice of Technical Writing,* a different book than Rickard's, but one still, notes Connors (1982), that included largely the rhetorical modes. Technical writing for Earle "was narrative, descriptive, expository, or directive" (p. 333).

Earle suggested earlier. Earle believed that ability to describe, applied to technical subjects, could inform an instructor as to the student's level of understanding, a concept
hinted at in the proceedings of the previous decade. He continued:

> In some of the most important parts of engineering work, such as designing, inventing,
> planning and organizing, the thinking has to be carried on with the objects present only
> as ideas. (p. 38)

This ability to translate the conceptual into the written, with audience considerations
and clarity of style, was the unique province of the engineer.

In another 1911 paper, presented by F.N. Raymond, Assistant Professor of English
at the University of Kansas, the topic of description again emerged, with Raymond
advocating lessons that emphasized "accurate and thorough observation, clear perception of purpose, and adaptation to purpose" (p. 50). This belief in the importance of
technical description was not only suggested by both Earle and Raymond, but reinforced in the ensuing SPEE discussion. One faculty member, Harwood Frost, by letter, praised Earle in particular for noting the importance of description, and Raymond
for his "attention to observation," because "no one can advance in learning without
some cultivation of his powers of observation" (p. 71).

THE ROLE OF FACULTY COOPERATION

Perhaps the greatest departure Earle made in his *radical experiment* was to suggest that
not only should new instructors serve a one-year apprenticeship with an established
English faculty member, but that the new instructor should also "be in actual contact
with engineering work" (p. 43) and engineering faculty. In fact, he believed that
"instructors in English in engineering schools should be trained engineers, and teachers of applied science certainly ought to be masters of good English" (p. 44). However,
he wistfully acknowledged, because either is unlikely, more faculty might consider
cooperating with one another as a solution to the problem.

In fact, Earle's call for greater cooperation between English and engineering faculty was a means for him to comment, however obliquely, on an enduring problem in
engineering colleges—many of the engineering faculty themselves either exhibited
poor English communication skills or, in fact, when those faculty discovered English-
related errors in their students' papers, did not note the problem, but joined in the cry
for better prepared students. The lengthy discussion that followed both Earle's and
Raymond's papers indicate that SPEE members, although disagreeing with some
points, in large measure agreed that students needed "an atmosphere of good English
for four years" (Gehring, p. 67).

Of course, underlying this call for cooperation among departments was the belief that
English was a tool common to all faculty members. In fact, wrote Professor Felix E.
Schelling (in response to Earle's and Raymond's papers), "English is our universal tool;
without English we can do nothing" (p. 75). This notion that English was not strictly the

province of English departments was even echoed in a July 1, 1915 letter, "English in College" to *The Nation* by a Cornell professor who decried the tendency to place English instruction solely on English faculty, particularly those faculty who taught in technical colleges: "In the next place it begets and encourages in the student mind the belief that good English is of use only in the English room, but does not count outside" (p. 174).

So great an issue did cooperation become that the SPEE requested a paper on the subject in 1913 by John T. Faig, a University of Cincinnati Professor of Mechanical Engineering. Certainly one reason for the requested paper may have been Faig's description of cooperative courses as *radical,* implying that interaction among the engineering departments was as much an issue as was the interaction among engineering and nonengineering departments. The SPEE may also have requested the paper to evaluate the usefulness of constructing courses that required interaction between engineers and English faculty, particularly in light of the lengthy discussions on English prior to 1913. Faig's description of cooperative courses certainly involved English faculty, who, in keeping with the previous point made here on the relevance of description to the engineering student, "criticize the descriptions, arrangement of matter and clearness of expression" (p. 99). The conclusion of Faig's piece, rather than emphasizing the great benefits to students, instead referred to the faculty, their *frank* weekly discussions, and the ability of cooperative courses to *energize* their work. Although the usefulness of this approach for students remains evident in Faig's piece, there exists also an underlying sense that such a course might also serve the faculty.

This is not to suggest that all four-year institutions experienced difficulties in getting the engineering and English faculty to cooperate. Evidence implies, however, that for many such institutions serious problems did exist, problems that certainly must have had an influence on the direction of technical writing. The problem was two-tiered. First, the wariness of some engineering faculty toward anything related to the humanities continued to plague many institutions, leaving those who taught English isolated. Second, and perhaps most importantly, many English faculty experienced a kind of self-imposed isolation due to their literary training. Engineering English was simply not what many of them wanted to do.

The first of the aforementioned problems—the wariness of engineering faculty toward English faculty—was perhaps best reflected in the lengthy 1911 discussion that followed Professors Earle and Raymond's papers. Harwood Frost, referring to an unnamed dean at a large engineering school, cited the following: "One of the most perplexing problems among our universities at present is getting any sort of connection or sympathy between the English departments and the engineering schools" (p. 68). The implications are worth considering. Why would such a lack of sympathy occur when few could argue the relevance of English instruction to general literacy? Did the engineering faculty simply mistrust the liberal arts education of the English faculty? Was there some sense that the mission of the technical college was being circumvented by humanistic training in English (particularly literature)? Did English faculty conduct themselves with an implicit attitude of superiority, thus exacerbating the already obvious divisions between engineering and English instructors? Although it is impossible to know with certainty the answer to any of these questions, probably some combination of actual problems and perceived difficulties contributed to a lack of sympathy among faculty at some schools. That a large proportion of English faculty at engineering schools assumed an air of superiority seems the least likely scenario. Although those teaching literature, in general, may have retained some semblance of a humanistic, elitist sense about their topic, the vast majority of those teaching composition or engineering English (a composition hybrid rapidly approaching a technical writing course), not only likely felt little superiority attached to their positions, but instead probably felt the sting of low status usually accorded *service* courses.

THE SECOND-CLASS STATUS OF ENGINEERING ENGLISH

Perhaps one of the greatest ironies in the move to technical writing as a discipline is that a first-rate English alternative for engineering students emerged almost in spite of its second-class status. Compounding the irony is the role of English faculty who virtually imposed this status on themselves. Many of these teachers found themselves teaching a subject for which they were not trained and looked down on by colleagues who were teaching what virtually all English faculty were trained to do—teach literature. Consider, for example, the relatively low status that composition instructors held teaching service courses. Transmute that service course into the hybrid engineering English course and you had faculty who were perceived, as Russell (1991) noted, "beneath the teaching of literature [and] beneath the teaching of engineering" (p. 122).

Perhaps what slowed technical writing's evolution even more, and certainly contributed to some of the lingering problems facing any engineering English in the curriculum, was the self-perception of many young faculty members who had, Connors (1982) noted, "no glory and no real chance for professional advancement in technical writing" (p. 335). Simply put, many of the faculty were not committed to teaching such a service course, which meant also that many of those same faculty were not actively pursuing professional development connected to their teaching; many were either attempting to find the means to teach literature or were using the service courses to improve their later chances at advancing to more literary pursuits. Remember, too, that the concept of technical writing, just an experiment at Tufts during roughly 1911–13, did not constitute a certain enough academic path upon which faculty members could base their careers.

Furthermore, why should these teachers embrace such an uncertain path when many, in and out of academic fields, were arguing persuasively for more culture in colleges and universities, particularly in light of the still-declining literacies of college graduates and the ever-increasing popularity of vocational or utilitarian curriculums? A September 26, 1912 *The Nation* editorial, "Practical Education," decried the call "for immediate utility as the chief end of education" (p. 278) as a direct reference to "the new, so-called practical subjects" (p. 279). The conflict—culture versus utility—is certainly not new and in many ways remains an issue even today as technical writing faculty still find themselves perceived, sometimes, as less important or less scholarly or even less legitimate than those who teach literature.

Even in this relatively negative atmosphere for the development of technical, practical writing, changes were imminent, even for those still favoring the failed status quo—an engineering English virtually devoid of context for the students. Change, the evidence would indicate, came slowly between Earle's paper in 1911 and the subsequent formation of the Committee on English in 1914, a committee charged with first evaluating the present system for English instruction and then recommending needed changes. Interestingly, during the three-year period between Earle's paper and the new Committee on English, the SPEE proceedings reveal little if any activity on the matter of engineering English. Were those three years a period of intense debate on the issue? Did the SPEE, because of Earle, decide finally to study English as a key curricular issue? The answer to these questions is, of course, impossible to know. But Earle's influence seems to have played a large role in the formation of the Committee on English, a committee that would report regularly on the status of English in an engineering curriculum, for Samuel Chandler Earle was appointed the first chair of the committee.

THE COMMITTEE ON ENGLISH: LOOKING FOR ANSWERS

The Committee on English was charged not only with evaluating the current role of English in the engineering curriculum, but also with establishing the climate for English both among English and engineering faculty. This poses two key questions: Why did the committee find it necessary to measure the climate, and how did the committee undertake such a measurement? Evidence provides a good answer to the second question. The first question is somewhat harder to answer.

The uncertain role of English in the engineering curriculum and the perception of that uncertainty by faculty help explain why the Committee on English felt constrained to evaluate climate. Simply put, the attitudes of both engineers and English faculty toward the teaching of English in a technological institution fostered a two-pronged mission problem. First, enough engineers still believed that English had no role in an engineering education. If they did acknowledge the role of English, it was to grant the subject *culture* status, further exacerbating the perceived distinctions that already existed between engineers and English faculty. Thus, it is possible that some engineering faculty (although certainly not all) believed English—the course and the faculty—might potentially subvert the mission of an engineering college. Second, and perhaps more insidious to the evolution of technical writing as a discipline, was the perception of the English faculty. Trained primarily to teach literature and the product of a liberal arts environment, English faculty either taught literature in engineering colleges to a less than receptive audience, or they taught composition (engineering English), a course that enjoyed only second-class status. Thus, it is similarly possible that a far larger number of English faculty perceived the unappreciated literature course or the engineering English course as a potential threat to the mission of an English (or liberal arts) education. In other words, engineers and English teachers were not just psychologically battling each other, but they were battling themselves.

In fact, C.W. Parks (1916), substituting for a very ill Samuel Chandler Earle, noted in the first report from the Committee on English:

> The attitudes reported as between the instructors in English and the instructors in other departments extend all the way from open hostility in a few cases to entirely sympathetic cooperation in others....Between the two extremes of hostility and sympathetic cooperation is a sort of mild tolerance of one another's existence, that is about as unfortunate as open hostility, because it represents indifference regarding the whole question. (p. 180)

The Committee, in order to evaluate the climate, mailed approximately 107 surveys to faculty at engineering colleges, asking questions that ranged from issues of cooperation and English promotion by other faculty to questions on the role of and relative importance of English. Many did not reply at all. To some questions, the Committee reported as few as 12 replies, a figure that makes identification of a consensus difficult. The first report of the Committee on English, frankly, did not offer the kind of encouragement faculty may have been seeking. However, one useful issue did emerge, which certainly must have pleased those teaching engineering English: salary.

By considering the previous comments on English faculty perceptions of themselves, coupled with the very low pay offered faculty to teach English in engineering schools, another factor emerges that contributed to the slow evolution of technical writing as an English discipline. The impetus to teach such a course existed on no discernible level. The job was ill perceived by engineering and English faculty alike, students largely rejected the discipline, the position held at best second-class status, and the pay was so far below the norm for faculty that there existed, simply, no real reason to care about the subject. Parks commented in his report that engineering English teachers do not aspire:

> to become teachers of engineering English; not to analyze their problems and understand them better; but to get a disagreeable job off their hands as quickly as possible, in order that they may bask in the sunshine of pure culture in some other more congenial department. (p. 182)

The earliest form of technical writing, it now seems clear, was a means to an end for many faculty, a chore to perform prior to rewards in literature. No wonder engineers mistrusted English faculty and the courses they taught. English faculty themselves could not embrace their own discipline in this form.

It was, ironically, the best and worst of times for engineering English. The fact that Samuel Chandler Earle, in 1911, presented such a strong argument and format for a prototypical technical writing course was clearly the signal many needed to take notice of writing and its potential. By 1914, though, Earle's plan was a hazy memory. The Committee on English, with its surveys and climate evaluations, was certainly a move in the right direction, but Professor William Kent noted in response to that first English Committee report that "the present status of teaching English to engineering students...is bad" (p. 183). So bad, in fact, that even the metaphors used to describe those teaching English implied second-class status. Professor W.T. Magruder, for example, called "English the scullery maid of our engineering college household" (p. 185). The first report of the Committee on English, finally, answered few questions, but the report did at least help to shape the direction of future questions.

Parks and the Committee on English established the first, broad question in 1917 when they surveyed teachers of English to determine (a) what kind of English instruction engineering students should have, and (b) what the aim of that instruction should be. Although the responses indicated a range of aims from a "guarantee against illiteracy" to "a cultural and recreational escape from the monotonous literalism of vocational study," the majority of respondents indicated "a growing tendency to give engineering students not general composition but instead some kind of instruction in technical writing" (p. 218). Earle, undoubtedly, had something to do with this very early and gradual shift toward English as a tool.

Although ill, Earle remained chair of the committee and certainly exerted some influence on the direction of the committee's recommendations. For example, the committee under Earle mailed 25 copies of a survey-style letter to engineering companies, asking the recipients to comment on a young employee's knowledge of English. Of 15 respondents, 8 felt that recently hired engineers did not have adequate English skills to perform his work. The cause? Overwhelmingly respondents pointed to the engineer's inability to construct reports that were logical, systematic, and grammatically correct. Clearly, if the real-world respondents in this survey are any indication, Earle was right when he argued so convincingly in 1911 for a different kind of English course. If ultimate success is measured according to the ability to excel in a profession, English had to become a larger factor in the life of an engineer.

1917–1919: TEXTBOOKS AND CULTURE'S LAST STAND

The one issue not really raised by the SPEE during most of this decade was the role of textbooks in teaching engineering English. Certainly, literary texts—novels, poems, and short stories—existed where literature was still taught, but textbooks written specifically for the engineering English instructor were largely limited to Rickard's book. This text, discussed in Chapter 2, was essentially a composition text with great emphasis on the rhetorical modes. In 1917, however, another text appeared on the engineering English market, and this book, because of its author, not only carried some weight, but also may have slowed the progress toward a technical writing course as the primary English course in an engineering student's curriculum.

English and Engineering, edited by MIT faculty member Frank Aydelotte (1923), was striking for two reasons. First, Aydelotte was a member of, and great contributor to, the SPEE Committee on English. Second, Aydelotte, himself an engineer, made no secret of favoring English as a cultural influence in the engineering curriculum. His text, largely a prose models reader of selections by everyone from Robert Louis

Stevenson to John Ruskin, emerged at the height of liberal culture consciousness at colleges all over the east coast. As noted in Chapter 2, liberal culturalists, according to James Berlin (1987), embraced an elitist approach to teaching writing, usually drawing on literature as a means to teach composition. Aydelotte, in his choice of selections and in his introduction, adopted a similar approach.

Connections between the two movements are apparent. First, Aydelotte taught at MIT and thus likely found himself exposed to the East Coast liberal arts milieu that supported a liberal culture approach in teaching writing. Second, Aydelotte, like the liberal culturalists, believed in two great divisions—literature and science. Third, like the liberal culturalists, his plan for teaching writing proved useful only to the most motivated, intellectually gifted, and well-prepared students.

Aydelotte stated in the introduction of his text that he believed that engineering students could best learn to write and think through exposure to literature, "once the connection between engineering and literature is made clear" (p. xiii). He perceived literature as a means to "train students to express themselves [and as a means] to furnish something of the liberal, humanizing, and broadening element which is more and more felt to be a necessary part of an engineering education" (p. xiv). Furthermore, as Connors (1982) pointed out, "On the east coast there grew up a movement led by Frank Aydelotte" (p. 334) [5] The size of this movement among educators is unclear, but an English professor from Rensselaer Polytechnic Institute, Ray Palmer Baker, may have aligned himself with Aydelotte by producing a 1919 text, *Engineering Education: Essays for English,* which presented a similar theme.

Baker's selections were not only arranged in a manner similar to Aydelotte's text, but also included some identical selections. Perhaps more telling, though, was his reference in the text's introduction to "the earlier phases of the debate between the champions of utility and culture in education" (p. vii), again, like Aydelotte, reducing education to two *camps.* Baker also mentioned Arnold, whose definition of culture, according to James Berlin, was to the liberal culturists "the best that has been thought and said" (p. 45). Although Aydelotte's text went into a second edition (1923), both books were clearly produced to maintain the English status quo at engineering colleges. As the decade closed, however, a third figure emerged who had as significant an impact on the engineering English field as did Samuel Chandler Earle—S.A. Harbarger.

1920: HARBARGER AND FREEDOM FROM THE CULTURE OBSESSION

The report from the Committee on English to the SPEE in 1919, the same year that Baker's book was published, discussed two methods of teaching engineering

---

[5] Aydelotte's East Coast movement coincided with the Liberal Culture movement, also developing on the East Coast. The aims of both movements are quite similar—literature as a means to advance the humanities.

English: one a final nod to Aydelotte's liberal culture approach, and the other foreshadowing Harbarger and the future of English in engineering colleges. In using the MIT English curriculum, Park and the rest of the English Committee inevitably turned to the "relation of science to literature," and the authors second term freshmen read "Arnold, Newman, Carlyle, and Ruskin" (p. 320). Clearly, Aydelotte's belief that literature was the means to English instruction endured. However, by the end of the report, Park and the rest of the committee presented a laboratory report outline, generated by engineering and English faculty, as a "practical suggestion" for teachers (p. 324). The outline (see Figure 3.1) represented a merger of technical and report writing concerns, concerns that would be manifest in Sada Harbarger's upcoming (1920) presentation to the SPEE.

Although laboratory reports differ widely in subject matter, the problem of organization is very similar in all of them. The outline given below represents a logical development of the subject as well as a convenient distribution of the material in the report of a test. With a few changes here and there to fit special conditions, this form will be found serviceable in the writing of nearly every kind of laboratory report.

1, *Object.*—This division should consist of a clear, full, and concise statement of the object, preferably in the form of a simple declarative sentence. Since the report is not written until after the test has been performed, the statement should be put in the past tense; e.g.,

"The object of this test was to determine the steam consumption of a Brownell 10 X 12 engine."

2. *Theory.*—This division should contain a general statement of the data to be obtained in such a test, together with the fundamental principles on which the test depends. For example, the second paragraph of the report indicated above might begin as follows:

"In testing for steam consumption, the chief data to be obtained are (a) the horsepower of the engine, and (b) the weight of the steam passing through the cylinder in a given time."

Where formulas are to be applied in the test, they should be given in this part of the report.

Since the discussion deals with general theory, this section of the report should be expressed in the present tense.

3. *Apparatus.*—The apparatus used should be described, with emphasis on new apparatus and on special devices used in the particular test in question. At the beginning of the section, the various pieces of apparatus should be enumerated. New and special pieces may then be described in detail.

In so far as this part of the report deals with particular apparatus, it should be put in the past tense; e.g.,

"The apparatus used in this test consisted of the following," etc.

4. *Procedure.*—This section contains an account of what was done in carrying out the successive parts of the test. Care should be taken to omit preliminary and non-essential operations, and to follow the actual order in which the work was done.

The narrative should be impersonal, and should be given in the past tense and the passive voice; e.g.,

"Readings were taken at intervals of five minutes," etc.

5. *Results.*—(a) Summary of results: Conclusions drawn from the data should be stated briefly and clearly. In some cases it may be desirable to compare them with results obtained in other tests. (b) Curves. (c) Sample calculations: These may be brief, but they should be typical of the mathematical processes involved. (d) Data: The data should be presented on special paper designed for the purpose, and should be arranged in tabular form and in parallel columns.

6. *Sketches.*

7. *Original Data.*—Rough notes taken during the test should be submitted from time to time as evidence of the accuracy with which observations were made. If a log book is kept, it will answer the purpose.

Respectfully submitted,

C. W. Park, *Chairman.*

Figure 3.1. TYPICAL OUTLINE FOR A LABORATORY REPORT.

S.A. Harbarger's (1920) presentation, "The Qualifications of the Teacher of English for Engineering Students," in many ways concluded the second decade of the 20th century by attempting to answer questions implicit throughout the decade—namely, the role of English and of English faculty in engineering education.[6] Her remarks, very much in keeping with Earle's (1911) proposal for a different kind of English course, provided the foundation for discussions about staffing English courses. It also diminished the belief that English was the last bastion of culture.

The role of the engineering English teacher, according to Harbarger, was threefold: to reinforce engineering principles through English instruction, to connect English to the future professional life of the engineer, and to view English as the link to professional and social success. She was not, however, arguing that literature was the means to those ends. Instead, Harbarger believed that engineers needed English more than they realized; the qualified teacher, therefore, would make that need known and would help the student to regard English as a tool. She wrote that "he [the instructor] associates English, therefore, with reality and finds an objective for his students' thoughts... He makes the connection of English with engineering apparent" (p. 302).

In addition, Harbarger called for teachers who wrote articles for technical journals in the belief that the reputation of English instructors with their students would rise if their work was published alongside the work of engineers. "The English instructor," she wrote, "has a splendid chance to demonstrate that he can practice what he preaches" (p. 303). Although Harbarger encouraged English faculty to write for technical journals in order to increase students' estimation of them, there seems little doubt that she also meant, at least implicitly, that perhaps engineering faculty as well might regard the English instructor differently if similarities rather than differences between the humanities and sciences were highlighted. Students and engineering faculty, stressed Harbarger, needed to see what English had in common with the sciences, not what it could do to improve the sciences.

If her presentation to the SPEE was any indication of Harbarger's commitment to engineering English education, her contribution to the English community became far more compelling in the next decade. In 1923, she would publish *English for Engineers,* a book that hinted at the future of technical writing. Of course, 1920 also marked the second edition of Rickard's text, *Technical Writing,* but the second edition really looked a great deal like the first, with emphasis on the modes and grammatical correctness. Harbarger's book, comparatively, was a

---

[6] Harbarger, Assistant Professor (Department of English) at Ohio State University, is described by Connors (1982) as "tough-minded and professionally determined" (p. 335), qualities that are implicit in her writing. Connors also suggests that her decision to use the initials S.A. (rather than Sada) in her published material was "perhaps because the publisher felt that many readers might resent a woman claiming to be able to teach technical writing" (p. 335).

departure from Rickard's and signaled some of the significant changes ahead in the gradual evolution of technical writing as an engineering discipline. Not only was Harbarger a member of the Committee on English, but both her academic suggestions and her textbook, along with recommendations made by Earle at the outset of this decade, set the tenor for discussion in the decade ahead—the synergistic formation of a discipline.

Chapter 4

1921–1930:
Teachers, Textbooks, and the Move to Cooperation

If the last decade was notable for the many questions that emerged on the role of English and English teachers in an engineering curriculum, the 1920s in America were notable for some attempts at answering those questions. Perhaps the most crucial question—Who should teach English courses to engineering students?—was answered in part by Sada Harbarger in her 1920 piece on "The Qualifications of the Teacher of English for Engineering Students," but the question of qualifications was one that preoccupied the Society for the Promotion of Engineering Education and the Committee on English and English faculty throughout the 1920s. As discussions on qualifications naturally led to qualifications for all faculty, the possibility of university-wide cooperation on English curricular matters arose. English teachers, who for so long felt isolated by the engineering faculty, began to encounter a greater degree of acceptance and indeed encouragement by their engineering colleagues. As this chapter demonstrates, interdependence contributed to the synergistic underpinnings of a technical writing curriculum.

ENGINEERING EDUCATION: SYSTEM OR MELANGE?

The issue of teacher training early in the 1920s was not limited to questions about English faculty; engineers themselves, as early as 1921, began to question the inconsistencies in their teaching styles and curricular requirements as well. In fact, the SPEE itself, a national clearinghouse for ideas and trends related to engineering education, came under fire in 1921 by W. H. Burr, a New York City consulting engineer. Writing on "Some Features of Engineering Education," Burr (1921) wondered if there was any stability and/or continuity in the field of engineering. He decried the state of engineering education, commenting that the "topsy turvy mixing of its [engineering's] elements as that now being exhibited in the engineering field [is] evidenced in no more conclusive manner than by the publications of this Society" (p. 65).

He believed that the failure to accord status to engineers resulted from the lack of continuity in programs, the irregular nature of specific requirements from school to school, rendering engineering education "simply a melange and not a system" (p. 67). Burr felt that the engineering curriculum itself was not working, and he went on to imply that the discipline had not evolved into what it hoped to become—an academic profession comparable to legal or medical pursuits. If, then, engineering was still being molded into a viable discipline, engineering English, at best still an afterthought in many four-year institutions, lagged even further behind. Indeed, engineering English either contributed to this lack of perceived curricular continuity, suffered as a result of it, or both. Clearly this was a growing problem for the engineering community, which called for careful investigation of engineering programs across the country.

Burr, in fact, concluded his paper by calling for a large-scale study of engineering education in America. His voice was only one of many that motivated the resulting report—the 1923 Wickenden study of engineering curricula—which attempted to analyze the current state of engineering education and propose a unified engineering curriculum. A notice in the November 23, 1923 *Science* announced the "Study of Engineering Education," funded in part by the Carnegie Corporation, which "set aside the sum of $108,000 for the purpose of making possible a study of engineering education" (p. 417). William E. Wickenden, assistant vice president of the American Telephone and Telegraph Company, was chosen to direct the project, certainly the most extensive investigation of engineering to date. Among the many areas to which Wickenden and his evaluation team turned their attention was the lack of adequate teacher training, with the role of the humanities in engineering education figuring prominently.

RECONSIDERING THE HUMANISTIC CONTEXT

Contained in the SPEE Proceedings for 1921 was the transcript of a rather brief exchange among several engineering faculty members from around the country that clearly points to the belief that engineering teachers—both old and new—lacked an adequate understanding of their students. Instructors may have had a thorough understanding of their discipline, but their ability to communicate with their students was perceived to be minimal, thus establishing a barrier between the components of a discipline and the humans engaging in either teaching or learning that discipline. This lack of the humanistic context negated the academic material that students encountered because students could not *relate* to the instructors. In fact, the transcript of the "Symposium on Training of Engineering Teachers" (1921), in general, related the belief of faculty that new engineers had to consider many factors in choosing a teaching career; most importantly, those new engineers needed to believe in their discipline so strongly that they desired to share their knowledge with others. In this, many new engineers were perceived to be lacking. Why? As G. C.

Anthony, a member of the symposium, expressed it, "I lay very much less stress upon the engineering teacher's knowledge of his profession than I do of his understanding of men" (p. 111). Without a humanistic context for engineering instruction, new and presumably future engineering faculty had no connection to their students. What robbed the engineering curriculum of this humanistic context was the separation of the cultural from the technical.

In his 1922 presentation, "The Cultural Element in Engineering Education," Professor W. H. Rayner pointed out that context for engineering was restored when teachers did not regard the technical and the cultural "as separate compartments in human life" (p. 155). Indeed, as we have seen throughout this study, the segregation of disciplines not only led to the displacement of English faculty in engineering institutions, but culminated in the debate over how to train engineering faculty to teach their own disciplines. Although many of these issues have been raised previously—the role of culture, the place of English, the qualifications for faculty—few of those broad queries resulted in little more than new questions. By 1922, those questions were being asked often enough and by educators and engineers alike, so that answers, or possible answers, began to surface. Not surprisingly, one potential answer to the problem of teacher training—all teachers—was to instill a belief that all disciplines are interrelated. The engineer, then, could embrace the humanistic as well as the technical. So, too, could the English teacher.

THE INTERDEPENDENCE OF ENGLISH AND ENGINEERING

The 1922 Report of Committee No. 12, English revealed not only that the first SPEE English conference had been held, but in listing the conclusions of those attending the conference, defined the movement of engineering English curriculum for the 1920s. The first ever English conference was called so that English faculty from around the country might meet and interact; it was also called so that those faculty could come to some consensus on the primary issues involving English and engineering teachers in the future. The English Committee condensed comments, concerns, and other topics into the following list of three items:

1. Our objective is identical with that of all teachers of English, i.e., to develop the student's ability to express himself orally and in writing with simple effectiveness; to enlarge his fund of ideas, and to increase his ability to classify those ideas; and finally, to create a sincere taste for discriminate reading and...standards of critical judgment.
2. Our method of approach must be to begin from and build on the professional interest of the student.
3. Our method of procedure must be to establish at all points the closest possible interrelationship between the instruction in English and that in professional subjects. (p.199)

These points are crucial to understanding not only the movement of the decade in English curricular matters, but also how technical writing in particular might have evolved from these beliefs. The first point—that all English teachers seek the same end—clearly positioned those teaching engineering English with all other English educators in the country. The second point—that the approach to English must draw on the students' professional goals—fully cemented the apparatus for English instruction in engineering colleges. The third point—that an interrelationship with other subjects advanced English instruction—most clearly established the foundation for a future technical writing curriculum.

The move to technical writing in an engineering curriculum in all likelihood evolved synergistically from this shift to curricular interdependence. Consider the two halves of *technical writing. Technical* implies the objective, the manipulation of data, objects, numbers, or otherwise linear presentations of material or ideas. *Writing,* on the other hand, implies the subjective, the manipulation of ideas, thoughts, intangibles. Putting the two together, up until this point, was naturally difficult because the inherent nature of each effectively negated the other. That is why for so many years—decades, in fact—students labored to make English relevant when it was not. P. B. McDonald, in his 1920 *English Journal* article, "Engineering English," noted this lack of relevance:

> The engineering colleges complain that the professors of English do not teach the kind of English that the engineering students require. Instead of training young scientists to see language as a wonderful and precise tool…the English professor too often has disgusted his class by pedantic analyses of dry essays. (p. 589)

Engineering English involved a process foreign to the student, in that the composition curriculum largely asked students to write themes on abstract concepts. Students, after leaving such a class, proceeded to a drafting or mechanical engineering class, where they were instructed to conceptualize what would become the tangible (e.g., a blueprint). Therefore, the class that should have facilitated their ability to conceptualize—English—instead emphasized conceptualization of the intangible, which engineering students rejected out of simple rebellion or, frankly, confusion.

Thus, the move to interdepartmental cooperation merged the subjective function of English with the objective apparatus of an engineering curriculum. What do engineers do primarily, both in their curriculum and in their careers? Engineers, wrote W. E. Wickenden in a 1927 *English Journal* piece, are "essentially visualizer[s]…[who] think in terms of pictures, diagrams, and charts" (p. 447). Because, Wickenden continued, so many engineering students are not proficient in either written or spoken English, many resort to "supplementary language-forms associated with visual rather than auditory perception, such as mathematical symbols and formulas, graphs, and drawings" (p. 447). The shift to *technical writing* as a byproduct of interdepartmental cooperation may well have been an acknowledgment of this point. Engineers engage in abstractions only to the extent that those abstract images are ultimately manifested first on the page and second in the form of an object, mechanism, or structure.

THE DEBATE OVER WRITING ASSESSMENT

If, though, the engineering and English faculty were to cooperate with better written expression as their joint goal, who, ultimately, should be responsible for the assessment of student work? Whether it was the intent of the June 21, 1922 SPEE Conference of Teachers of English (chaired by J. Raleigh Nelson) to deal with the question, the question was nevertheless discussed at length. Faculty, both engineering and English, from a variety of institutions around the country shared their ideas on English curricular changes and/or greater academic cooperation, culminating in a sometimes heated discussion of the problem of assessment. If *technical* and *writing* should somehow merge via engineering and English cooperation, who graded the students' work—engineering or English faculty? Naturally, faculty from both disciplines had concerns about the other.

Just as the call for academic cooperation certainly facilitated the move to a technical writing curriculum, so the debate over assessment of student work must have signaled ultimately the need for engineering and English faculty with adequate education and/or background in both technical and humanistic fields. For example, Pittsburgh engineering Professor W. E. Mott, when asked if English instructors who cooperated with engineers had difficulty understanding the technicality of a student paper, responded: "I expect they do. My mind would not work very well criticizing the English if I were capable of it, without knowing what the man was writing about" (cited in Nelson, 1922, p. 264).

Conference Chair J. Raleigh Nelson (an English instructor) responded:

> I teach the course in engineering reports to seniors. I do not believe anybody was ever born to begin more poorly adapted to do this work than I, because I had no technical training, or no particular taste for engineering. It was a very big cross to me. I took it up as a consecrated cross and I bore it bravely. I made myself think it was necessary. I know I could not read those papers as well as an engineer could have done it but I have been conscientious about it and I am reaping my reward. (p. 264)

Nelson wondered, finally, if his general lack of training might not make him a better audience for a technical paper. The ensuing discussion made it very clear that consensus would prove elusive. Professor Seavey, from Tufts, called his university "one big English Department.... We are not segregated—because each man is contributing something to the other" (p. 270). Hardy Cross, another English teacher, believed that "engineers had better keep their hands off English, and the English instructors better keep their hands off engineering" (p. 277).

This general lack of consensus as to who should evaluate written material and why raises, of course, the issue again of whether written expression can be separated from the technical information that facilitated the writing. This, I believe, is an artificial issue related to the continuing problem of poor grammar and mechanics in the writing of many engineering students. Can, for example, an engineering faculty member adequately locate the site of mechanical errors? Can English faculty, similarly, locate and note errors related to technical (engineering) correctness? This preoccupation with mechanical correctness would slow down the momentum of the cooperative education movement toward a coherent technical writing course, although some English faculty and engineers would continue to cooperate. Unfortunately, the speakers from several universities showed that cooperation sometimes took the form of proficiency *policing* designed to prevent the engineer with poor skills from graduating.

PROFICIENCY POLICING:
THE OTHER SIDE OF COOPERATION

Proficiency policing, first outlined in the SPEE's 1922 Conference of Teachers of English by the University of Illinois (Nelson, 1922), clearly demonstrated two enduring preoccupations related to engineering English. First, engineering educators, in devising a method of this sort, sought to measure the effectiveness of a subjective discipline solely by objective means. Second, engineers wanted tangible evidence of the success or lack of success of their students in all areas, and mechanical proficiency permitted this much more handily than did an essay exam that would have relied heavily on the *judgment* of English instructors. At the University of Illinois, for example, students were at the mercy of any instructor who reported them for poor writing skills. If reported, students' work was evaluated by "the Committee on Students' English...composed of a representative from each of the colleges of the University" (p. 266). This system, on the surface, encouraged a cooperative atmosphere at the university because (a) any instructor could report any student, and (b) the system appeared to legitimize the role of the English faculty.

What this cooperative evaluation system engendered instead was the further alienation of English from the engineering student's sphere of relevance. The approach to English under this model was virtually penal, for students never knew from whom they might receive notification of their poor mechanical English skills or, indeed, whether they might even graduate. Iowa State professor A. Starbuck (1924a), in a similar move, described a plan in 1924 whereby:

> Any instructor who observes poor English in the work of any student may consult the departmental English adviser, and together they may decide what action to take. This action may range from merely cautioning the student, to referring him to the Committee on Students' English for additional work. (p. 339)

Similar plans existed elsewhere, and although the purveyors of these plans described the benefits of increased cooperation when engineering and English faculty worked together to decide a student's fate, the English faculty only cemented their role as chief tormenters of engineering students, for they could, albeit in a committee environment, delay students' graduation dates. Fortunately, however, cooperation as a means to penalize engineering students ultimately gave way to something more viable for student and teacher alike. Cooperation in its ideal form emerged by middecade as a concept that permitted and promoted curricular change toward improving the engineer for the workplace.

RECONCEPTUALIZING COOPERATION'S FORM

As the first Conference of Teachers of English came to a close, two important points emerged. First, engineering as well as English faculty attended the conference,[1] clearly indicating the interest of both groups in discussing the issues surrounding English education. Second, Chairman Nelson, in his last statement to the gathered group, hinted at the future attention English would receive, not only by the SPEE, but by all engineering educators: "Our Committee is going to have a space in the *Bulletin* every month. We are going to keep that thing red hot with discussions of the problems of English teaching" (p. 282). Furthermore, English became a growing concern for the Society with the regular English column. The remainder of the decade proved to be a period of considerable growth in the shift to a technical writing curriculum.

The second Conference of Teachers of English (1923) returned to the issue of cooperation, as well as attempts at achieving full interdependence between English and engineering faculty. However, underlying the continued problems associated with full cooperation were answers, which became increasingly apparent as the decade evolved. Professor Baker (of RPI) stressed that at some engineering colleges, English teachers still "lacked enthusiasm for their task, and...saw no future in it [often because of] inadequate salaries...and also [because] English had never been made an integral part of the engineering course" (pp. 250–251). In fact, he continued, many English faculty teaching senior courses in English still held instructor rank. Conference members wondered why this was still so when the benefits of English were so apparent, not to mention the benefits resulting from the full integration of English faculty with engineering faculty. The answer appears to reside in the willingness of both faculty and curriculum reformers to reconceptualize the role of cooperation.

Cooperation, up until 1923, had not worked to its full capacity because the concept of *cooperating* was misassociated somehow with the notion of *separate but equal,* thus missing the vital element in full cooperation for the advancement of engineering students. Generally, faculty cooperated with each other only insofar as the lines dividing curricula were not blurred. Engineers and English faculty could not even agree, ultimately, on who would evaluate the papers because both groups were placing emphasis on correctness—mechanical English or technical engineering correctness. Synergy had not yet occurred. Engineers still maintained the objective, technical status quo, whereas English faculty still bore the weight of culture as well as responsibility for grammatical skills. For engineering students to assimilate the full benefit of a reconceptualized cooperation, (a) engineering faculty would have to assume some responsibility for incorporating enculturating material into their courses, and (b) English faculty would have to reconceive their curriculum by deemphasizing the subjective and literary, reemphasizing instead written expression grounded in technical expertise.

---

[1] In fact, of the 33 in attendance, 19 were English faculty. Clearly, as one faculty member put it, that nearly half were engineers indicated that "a good many of us who are not teaching English at all are much interested in the subject" (p. 280).

RECONSIDERING THE ROLE OF COMPOSITION

The move to written expression grounded in the technical dominated both the SPEE proceedings on English, as well as emerging engineering English textbooks, for the rest of the decade. In order to promote technical written expression in the milieu of cooperation, however, English educators had yet to consider the role of composition in the curriculum. Literature, although still a requirement for many engineering students, was fading in importance by the 1920s, whereas English instruction in writing gradually came to be perceived as a tool equal in importance to technical training and certainly more than just a means to culture or liberal arts tradition. Composition remained a requirement for most engineering students, but as educators began to consider good written and oral skills crucial to future workplace success, composition's effectiveness was called into question. As mentioned earlier, composition coursework largely involved students' responses to abstract concepts or ideas. If the English teacher, in a spirit of cooperation, assumed more responsibility for the written equivalent of technical ability, then themes on citizenship, courage, honor, and so on, would become less and less relevant.

Bradley and Merwin Roe Stoughton, both of Lehigh University, wrote in their 1924 presentation, "Education in English for Engineering Students," that "exercise in composition is unwise training for prospective engineers." Because even though students should be spending 95% of their time "learning the fundamental principles of *how* to write clearly and concisely, [they strain and exhaust themselves instead] to think up *what* to write about a given subject" (p. 143). In other words, composition fundamentally failed the engineering student who became preoccupied with choosing a topic suitable to the composition classroom, when students should be preoccupied with the structure and process of writing, which could occur naturally with topics focused on the technical. Thus, rhetorical modes introduced in composition class—such as description, definition, and even process—could be retranslated for a specialized, more technical audience. The Stoughtons, in fact, suggested that a composition *atmosphere* might benefit a student "by giving him all his data and teaching him the principles of preparing it for both verbal and written presentation" (p. 145).

Clearly, students needed the basics in composing a written effort, and the composition classroom could provide that. However, note the emphasis on *data* and *presentation,* two elements not specifically associated with composition. Writing for engineers had to transcend the rhetorical modes, thesis/support structures, and a general subjective framework. Students needed to incorporate data into their writing as well as consider how best to present that data for a specific audience.

The gradual move in the 1920s toward greater English/engineering cooperation, began to establish technical writing as a composition hybrid. Although technical writing would certainly evolve into something far more complex than simply an offshoot of another discipline, there exists little doubt that the course, as it was originally conceived, grew out of a general frustration with the effectiveness of standard composition instruction in an engineering environment.

The *Ames Narratives,* presented in 1924 as part of a series on "Selling English at Iowa State College" (Starbuck, 1924b), clearly demonstrated the manner in which standard composition devices were reconceptualized for engineering students. The *Ames Narratives* were, literally, narratives written specifically for engineering students to use because those students

> have been trained to believe that ivy climbing over a wall is beautiful...and golden sunsets are beautiful, but that everything else is drab and commonplace. [They have] never been shown that there is as much beauty and romance within a powerful locomotive whirling its great train across a continent. (p. 482)

The development of these narratives reinforces the point regarding the effectiveness of composition. Engineering schools and universities were becoming increasingly frustrated with (a) the continued lack of written skills by engineering students and (b) a coherent plan to remedy the problem. The *Ames Narratives* certainly did not go on to become the remedy, but as yet another attempt to solve the problem, they pointed to the gradual move toward rethinking the one English class all engineering students faced—composition. From middecade on, English faculty in the SPEE embraced this issue of change and suggested some definitive measures for the future.

WHO SHALL BE THE TEACHER?

By the end of the 1924 gathering of the SPEE, the Committee on English established a list of conclusions drawn from faculty and administrative letters, which ended with one of the key questions of the decade—"From what source shall we recruit our staff?" (p. 561). A condensed version of the list (as follows) clearly indicates that questions on the relevance of English in the curriculum have been largely replaced by questions about the person who will teach the English class:

1. There is now no longer any need to urge the importance of training in English for prospective engineers.
2. There is almost universal recognition, that, while the objective is the same and the subject matter the same, the problems of teaching English to engineering students is a real problem, and the students' interests form a strategic point of attack, and a natural motivation for such work.
3. The requirements in English have doubled in the past ten years, and in many schools are now approaching the possible maximum.
4. The need for more perfect coordination of the work in English with the work in technical subjects is being increasingly recognized.
5. The distribution of the responsibility for training the student to write and speak as an educated man should, is being more fairly placed upon the entire engineering faculty to the relief of an overburdened staff of teachers of English, and to the greatly

increased efficiency of the work.

6. There is an increased tendency to teach English to students in special sections, to put such sections in charge of teachers specially interested and specially fitted for this work, and a rapid increase in the number of schools with a separate set of teachers to conduct these classes.

7. The teacher is coming to be recognized as the supreme factor in this problem. (Committee on English, 1924, pp. 560–561)

That the teacher became a special concern was certainly due in part to several trends discussed thus far, including the role of cooperation in engineering English education. Essentially, the qualifications of this English teacher who may not necessarily teach literature and, indeed, may not ultimately teach composition either, shifted the emphasis of curricular discussions away from the engineering English course itself to the teacher. The 1925 SPEE Committee on English took up this issue by preparing the outline shown in Figure 4.1 for those attending the conference to use as means to generate questions and/or comments.

Two points are worth noting here. First, the Committee on English (1925) regard-

I. Classification of Freshmen.
  A. On entrance.
    1. By examination.
    2. By week of test work.
  B. At the end of a probational period.
    1. Two weeks.
    2. Six weeks.
    3. Eight weeks.
  C. After postponement of English I.
    1. To the second semester.
    2. To the second year.

II. Supplementary Work.
  A. To precede English 1—Sub-freshman work.
  B. To accompany English 1—Supplementary work.
  C. To follow English I—Make-up work.

III. Courses in English Composition.
  A. Required courses.
    1. English 1. 2. Follow-up work.
  B. Elective courses.
    1. General.
    2. Specialized courses.
    3. Business English, Commercial Correspondence. Reports Scientific Papers, Technical Journalism.

IV. Courses in Literature.
  A. Systematic courses in the history of literature.
    1. English
    2. American.
  B. Courses in the appreciative and critical study of modern literature.
    1. General introductory courses.
    2. Foreign literature courses in English.
    3. Special courses in the drama, novel, short story, and scientific literature.
  C. Supplementary reading lists.

V. Combination Courses.

  A. English and History.
  B. English and Economics.
  C. English and Government.
  D. English and Logic.

VI. Plans for Checking Students' Careless Habits of Expression.

VIII. Plans for Securing Cooperation of Entire Family.

VIII. Senior English Courses.

FIGURE 4.1. SURVEY OF THE STATUS OF THE TEACHING OF ENGLISH

ed any contributions on English curricular matters and/or faculty qualifications "a contribution to the work Mr. Wickenden has undertaken" (p. 650).[2] The Wickenden study, mentioned earlier in this chapter, was one of the largest studies ever undertaken of engineering college curricula in this country. Second, and perhaps most important, this outline hints at, although never directly addresses, another engineering English development gaining momentum in this country—the rapidly growing number of new engineering English textbooks available to faculty. The books available to teachers during the 1920s indicated not only what curriculum reformers perceived to be important for engineers, but also in what directions the textbooks might be leading teachers.

TEXTBOOKS AND THE ENGINEERING ENGLISH TEACHER

If the SPEE in 1924 determined that English writing instruction for engineers was a given, then aside from the routine literary texts and standard rhetorics, the books available to teachers of engineering English—the rudimentary, early technical writing course—reveal a great deal about the movement and direction of the course as, naturally, the books were written by educators themselves. Of the books written during the 1920s, three are worth special consideration, not only because of the differences in style and content, but most importantly because, for the many teachers who *inherited* an engineering English course, these books constituted, undoubtedly, the foundation of their instruction. As a result, the books, as much as the teachers using them, certainly guided the course of technical writing in America.[3]

T. A. Rickard's book, *Technical Writing,* appeared in its second edition in 1920, and although the book reflected some content changes from its 1908 first edition, it remained a book largely concerned with usage and mechanical correctness. In fact, in many ways it most resembled the equivalent of our contemporary composition or technical writing handbooks, with sections on clarity, punctuation, jargon, and so on. This book, however, went to a third printing in 1931, and clearly enjoyed an audience of those educators who still remained preoccupied with mechanical tasks such as spelling, sentence construction, and punctuation.

In 1923, Sada Harbarger published *English for Engineers,* a text that, unlike Rickard's, espoused a kind of workplace or "real-world" philosophy, an appeal to engineering students that they might embrace the relevance of English in their ultimate pro-

---

[2] The authors of the two-volume *Report on the Investigation of Engineering Education*, which provides results on an investigation spanning the years 1923–29, ultimately believed "in the choice of humanistic studies [though] the primary criterion is not one of intrinsic cultural or intellectual values, nor one of narrow utility, but that of functional relation to engineering" (p. 148).

[3] Furthermore, as Connors (1986) noted in his "Textbooks and the Evolution of the Discipline," textbooks can go from "servants to masters" (p. 180). For inexperienced engineering English teachers, this was likely the case.

fessional success.[4] Certainly, Rickard provided students with an introduction that spoke to the virtues of good English, but Rickard's tone was decidedly academic and therefore somewhat distancing. Harbarger's first chapter, "Professional Prestige and English," very clearly linked job-related success with English skills, a much more direct attempt to pull the student into the study of English.

Harbarger, in fact, spent three chapters convincing students that good English skills—and specifically good technical writing skills—made them more marketable. Although this may, on the surface, appear to be an appeal to the most vulnerable aspect of a student's psyche (ultimate financial success), Harbarger was really reflecting the consensus of the period. English had to be practical and work-related to be relevant to very busy students. In fact, before Harbarger moved on to discussions of letters and reports, she made very clear to students that ultimately they were the salesmen of their own ideas, and English represented the best means to sell them:

> To the employer of engineers in the large industries, skill in English is one of the specifi-
> cations for the employees who are, in one way or another, to be salesmen whether they
> are in the sales department or in the factory. (p. 5)

Finally, in 1925, Sam F. Trelease and Emma S. Yule published *Preparation of Scientific and Technical Papers* (1925/1937), a book, according to Robert Connors (1982), that was "immensely popular and influential" (p. 336). Connors believes the popularity was due in part to the book's narrow focus, but certainly the book's success must have been equally tied to its process orientation. Trelease and Yule focused the text on the construction and writing of technical papers and reports, but they also used the language of scientific process with which engineers were certainly familiar. Many of the chapters were laid out according to *steps,* with numbered instructions and detailed outlines accompanying them. If Harbarger, in her text, connected English with future professional success, then Trelease and Yule opted to attach English instruction to scientific and technical process, another way of attracting the engineering student.

Implicit, though, in all three of these texts is the underlying notion that the books must either reflect current concerns for the still general lack of good English skills in engineers (as Rickard did), or they must espouse a philosophy relevant to the engineer (as Harbarger did), or they must utilize the language and methods of science and technology (as Trelease and Yule did). Although other texts existed during the 1920s, these three provide a good sense of the movement and direction of the discipline.

THE UNREALISTIC SEPARATION OF FORM AND CONTENT

If textbooks provided an indication of the classroom habits of engineering English teachers, then a survey of those same teachers to find out what and how they were

---

[4] Harbarger published the second edition in 1928 with McGraw Hill Publishers.

teaching students would provide even more information. The SPEE Committee on English, as Connors also noted, loved a survey. In 1925, in their regular feature, *The English Department,* J. Raleigh Nelson printed a questionnaire sent to teachers of English in engineering colleges (see Figure 4.2).

The SPEE committee was very interested in how much cooperation existed between the English and the engineering departments, as well as in the pedagogical tools teachers used in the classroom. Although the Committee did not intend to endorse cooperation, implicit in the questions and the summary that followed responses was a belief in the usefulness of cooperation. The responses, though, did not indicate that cooperation was favored by all. For example, although 91% of respondents answered "yes" to question IIa ("would cooperatively teaching English promote the habitual use of correct English?"), only 43% answered "yes" to question IIc—"is cooperation feasible?"

The Committee isolated four reasons why English teachers perceived that engi-

I. Will you please explain any plan that you may have in use for the cooperative teaching of English composition; that is, a plan that encourages students to use correct English in all classes and not merely when they are in a class in English composition?

Is the plan used in the college of engineering exclusively?

In which other college or colleges is it used?

(The following questions are given as suggestions for topics to be covered in the discussion. *If you have not time enough to write a lengthy explanation, "yes" or "no" answers after each question will be appreciated.*)

a. Does the department of English grade papers written in other departments?

b. Do instructors in English assist in preparing papers (reports, for instance) for other departments?

c. Is the grade on papers written in other departments affected in any way by the grade given or by the recommendation made by the department of English?

d. Even if the grade is not affected, are students required to correct or rewrite papers in order to meet a certain standard?

e. Are instructors in other departments expected to deduct a certain amount for faulty use of English in papers written for their departments and not inspected or graded by the department of English?

f. Have you any definitely specified minimum requirements to guide instructors in checking errors

in English? (If you have, will you please submit a copy?)

g. Has the plan that you are using any marked weaknesses?

h. Have you statistics that show the correlation between grades given papers by /an instructor in a technical subject and grades given the same papers by an instructor in English? The grades given these papers by the instructor in the technical subject should be based upon content only, and not upon correctness of expression. The object is to find out if students who make high grades in their technical subjects are generally careful in their use of English? (If you have these statistics, will you please present them?)

II. Whether or not you have in use a plan of cooperative teaching of English composition, will you please state frankly what you think of the plan?

(The following questions are given as suggestions for topics to be covered in the discussion. *If you have not time to write a lengthy discussion, brief answers written after each question will be appreciated).*

a. Would a cooperative plan of teaching English composition, as indicated by the questions under topic I, promote the habitual use of correct English?

b. Do you think that a plan of this sort is fundamental; that is, can the best results in teaching English composition be obtained without it?

c. Is the plan feasible?

FIGURE 4.2. SURVEY ON THE VALUE OF INTERDEPARTMENTAL COOPERATION

neering/English cooperation might not work:

1. Instructors in non-English subjects lack accurate knowledge of good English to be able to cooperate intelligently.
2. Departments of English have neither sufficient help nor time to undertake the extra work that would be necessary.
3. Instructors in non-English subjects have not enough interest in good English to promote correct usage, or else they do not realize the importance of good English.
4. The burden of the work would fall upon the department of English. (p. 329)

Underlying, of course, all these conclusions is the sense that cooperation presupposes the separation of form and content. That is, to *cooperate* means that the engineer is qualified to assess the technical aspects of the paper, whereas the English teacher is qualified to assess the mechanical correctness of the same effort. The additional comments, which followed these four conclusions, indicated no real consensus, either, on the issue of cooperation. One English faculty member saw the value in cooperation of this sort, whereas another believed that "instructors in English are to be merely clerks to mark spelling and grammar" (p. 331). Cooperation of this kind—according to discipline specificity—undermined the nature of working together in the first place because of the continued insistence on maintaining separate (and presumably equal) bases for evaluation. To separate form and content in technical writing was virtually impossible because one aspect served the other. Form and content had to come together for technical writing to continue evolving.

INTEGRATED COOPERATION: TWO UNIVERSITY MODELS

At the University of Cincinnati and the University of North Carolina, such models did exist, and the underlying principles of both programs contributed another piece to the still evolving technical writing puzzle. Clyde W. Park (1925), in his "Details of Cooperation in the Teaching of English to Engineering Students," outlined the University of Cincinnati plan. English teachers evaluated the *whole* report written by a student, but with one difference—the engineering faculty assigned the topics. Although this still did not constitute the full incorporation of technical writing as a fully recognized English discipline, it did constitute another important move in that direction. In fact, in his description of the University of Cincinnati cooperative plan, Park also made clear that such an option was not only as fully viable as any composition component, but "better than the usual themes":

> They are real compositions, based largely on first-hand knowledge of the subject matter; they deal with material in which the student is interested and about which he has done voluntary, not perfunctory thinking; and they are written for a definite purpose. In their preparation, language is used as a means to an end, and not as an end in itself. (p. 399)

This plan, although certainly not perfect, clearly suggested that (a) cooperation in certain contexts accelerated the steady growth of what would become a technical writing curriculum; and (b) technical writing would, eventually, grow out of and replace composition as the most relevant English experience for engineering students. To take this cooperative initiative one step further, Louis B. Wright presented to the SPEE in 1926 the results of a similar plan at the University of North Carolina.

At the University of North Carolina School of Engineering, the engineering and English faculty redefined cooperation by arranging a course in which students "receive both technical and English credit for the same work" (p. 262). Students in this unique plan attended technical lectures and field trips, then wrote reports for their English component and spent, in conference each week, a predetermined amount of time with the English teacher who critiqued the technical work. Not only did the plan work, noted Wright, but, most importantly:

> however much one may emphasize the value of English as a cultural study, the average engineering student regards it as a purely utilitarian course, and he does better work when he sees its practical application. Far from destroying the individuality of English, the North Carolina plan of coordination turns the field of the student's greatest interest into the channel of study of English. (p. 264)

Although A. C. Howell, writing a year later in 1927, described the University of North Carolina "technical faculty and the division of Engineering English work[ing] in perfect harmony" (p. 621), the two university plans detailed in the SPEE proceedings did not solve the problem of English instruction for engineering students. Rather, they pointed to the role of unique cooperative plans in the gradual emergence of a technical writing curriculum.

WHAT WILL BE THE QUALIFICATION OF THE TEACHER?

As the decade gradually closed, the Committee on English notes and the papers presented before the SPEE indicated that despite greater English/engineering cooperation, one problem raised early in the decade still had not been settled: What constituted a "qualified" engineering English teacher? In fact, the 1927 Report of the Committee on English described the need for qualified teachers "an emergency" because the "demand seemed imminent for teachers qualified to give the engineering colleges the quantity and quality of English teaching likely to be called for in the reconstructed curricula" (p. 109).

The "reconstructed" curricula referred to the new emphasis on technical reports, construction of technical documents related to engineering projects, and letters and other forms of correspondence related to professional life. Clearly those trained for careers in literature and even composition required either retraining, further education, or at least a willingness to enter into the teaching of a discipline so new and so relatively unexplored. Apparently textbooks alone would not be sufficient for this reconceived English curricular specialty.

Teachers, the proceedings indicate, remained the potential problem in a fully cooperative and integrated environment because (a) so many still obtained degrees in literary work, and (b) so many of those teachers still perceived the work of English teachers in engineering schools to be little more than drudgery. During the 1930s, these questions of teacher qualifications, teacher satisfaction, and continued negative teacher perceptions of engineering English courses all pointed to one of the final pieces in the technical writing puzzle that remained to be resolved. Engineering colleges recognized and through cooperation were beginning to encourage English courses with a practical, technical basis. Textbooks, written largely by educators, followed suit with a variety of formats and styles to fit an evolving, *reconstructed* curriculum. In the 1930s, curriculum reformers had to consider not only the qualifications of the teachers who would teach this reconstructed English class, but the means to make the discipline attractive, challenging, and inspiring to this new kind of English teacher.

Chapter 5

1931–1940:
Composition, Technical Writing and the Site of Teacher Training

As the 1930s unfolded in American engineering education, the relevance of English in an engineering curriculum was no longer the big question. Replacing that was possibly a larger question: If English was standard practice for all engineering students, why were the students, the English teachers, and educators in general so dissatisfied with the results? The question proved somewhat more difficult to answer, although answers emerged in two engineering English trends that predominated the 1930s. First, teacher training, and especially the site of teacher training, took on greater significance. Second, the increasingly distinctive nature of the engineering English course caused educators to consider the discipline of technical writing as distinct and separate from the current composition requirements. This, plus the suggestion that engineering English teachers should be trained to teach differently than traditional liberal arts graduates, prepared the way for a new discipline.

As Connors (1982) noted in his history of technical writing, the 1930s were not only the period of the Depression, but the early 1930s "were not a happy time for engineering English teachers" either (p. 337). This was due to a number of factors, some mentioned in previous chapters. First, teachers of engineering English, particularly those teaching the course at an Arts and Sciences school in a four-year university, enjoyed little respect within their departments (populated by literature faculty), and even less respect from students.[1] Second, although teaching composition, a service course, was distasteful to many English faculty,

---

[1] Connors (1982), in fact, noted that students regularly regarded their English teachers as "effeminate," with one even calling his teacher a "budding pinko" (p. 337).

being relegated to engineering English meant (a) one was generally an inexperienced, young teacher, and (b) that one's chances for advancement beyond the instructor or assistant level were slim.

Needless to say, a good deal of the problem educators still perceived in English course offerings resided in the condition of the teacher. Simply put, no one was happy. To change this dissatisfaction in young teachers, those promoting engineering reform had to look closely at the reason for this dissatisfaction. Was this a systemic problem, a lack of consideration in the English departments? Or did this problem stem from the continued mistrust some engineers felt toward their more humanistically-oriented colleagues? The answer, of course, is not simply a *yes* to one query and a *no* to another. The answer, as the proceedings from the SPEE for this decade point out, stemmed from a very close analysis of two key concerns: the English course itself, and the background of the person teaching it.

UPPER DIVISION ENGLISH AS A POTENTIAL SOLUTION

As Connors (1982) noted, there remained a general sense of unhappiness with English courses in the early 1930s, so the SPEE spent considerable time analyzing the perceived problems, settling occasionally on a solution that not only worked in the short term, but that often contributed later to important pedagogical practice. For example, as early as 1931, members of the SPEE began to look at the logic of postponing one or more of the required English courses until later in students' academic careers because so many students seemed to indicate a greater appreciation for English by their junior or senior year. This is not surprising. By the third or fourth year of an engineering curriculum, most students faced not only report writing, but many may have worked as summer interns, experiencing firsthand the writing requirements of the real world.

This plan to postpone the English writing requirement provided a potential solution to the perceived English problem because, logically, students by the third or fourth year simply would have more information and education and thus more to write about. A.V. Hall (1931), in his "English as an Essential Part of the Engineering Curriculum," reinforced this notion when he wrote:

> The apparent waste of time involved in attempting to teach students who were not as yet interested and who had few ideas to express.... By the time men have had two years or so of technical training, they are prepared to use their intellectual powers upon a subject for which they now feel a real need. (pp. 416–417)

James W. Graham and William H. Barton, Jr. (1931) echoed Hall's belief in their "Practical English Course for Senior Engineers." The authors perceived "a vicious circle of indifference and antagonism" in both engineering students and English faculty because English "is rarely given beyond the sophomore year" (p. 604). Although postponement alone did not seem especially important in the larger scheme of engineering

English problems, the underlying contribution to the evolution of technical writing was inherent in delaying the English requirement.

Importantly, delaying the writing requirement until the junior or senior year of a student's program meant that the English faculty member teaching what was, in effect, an upper-level writing class required a greater background in English to challenge the more experienced students. Perhaps more importantly, the class, in all likelihood, was a technical writing course or some hybrid of it because postponement provided a means to allow third- or fourth-year students to bring more of their relevant engineering experience to bear in the writing process. Indeed, wrote Ray Palmer Baker (1932), "undergraduates...soon sense the attitude of their instructors. Above all they are alert to gradations of rank" (p. 290). Baker's belief was well substantiated by the responses of students themselves.

In 1932, as well, Robert I. Rees, Assistant Vice President of American Telephone and Telegraph, initiated a survey of 400 engineering students, recently graduated and employed by Bell System Companies. Responses to Rees's questions indicated a great desire by many of the students to take their English courses later in their academic programs. Rees (1932) noted:

> a large majority of men said the English seemed of little importance during the freshmen year; that by the senior years its importance had become apparent, and they now feel that is was very important indeed. (p. 482)

Indeed, Rees concluded, many "felt that English should be taught during the entire four years and that, if necessary to accomplish this, the engineering course should be lengthened" (p. 482).

Similarly, the 1932 SPEE Report on the Conference of English reinforced this belief in delaying the English writing requirement. C. E. Magnusson, English professor at the University of Washington, noted in the Report that placing the English requirement "in the freshman year automatically erects a barrier to satisfactory instruction" (p. 793). Recalling a freshman who did not want to waste his time on "that stuff" (English), Magnusson later noted that by the junior year, that same student requested composition because "this summer I had to write a report on some engineering work and my chief could not make head or tail of it. It is time I learned to express what I have to say" (p. 793).

Although this suggestion to postpone engineering English writing courses, at first glance, appears a simplistic solution to a far larger problem, the argument to postpone perhaps inadvertently reinforced the gradual move to technical writing as a discipline. The English course that both educators and apparently a good many students felt should be delayed was the writing course—namely, English composition. Why? Mostly because the composition course for engineers contained *components*, unlike the *straight* composition course that might be offered at four-year liberal arts institutions. The engineering composition course contained standard material (rhetoric, modes of discourse, drill work, etc.), but *technical writing* (mostly report, proposal, and letter writing), as well as journalism, were features of composition in many engineering English classes. In fact, A. M. Fountain (1938), in his definitive study, *A Study of*

Courses in Technical Writing, noted that the composition requirements in 117 schools of engineering that held institutional memberships in the SPEE were as follows, the units being quarter hours (p. 42):

elementary composition	7.40
advanced composition	.59
technical writing	1.40
business English	.11
journalism	.03

Clearly, technical writing was a part of composition, although this trend would change gradually.

A QUALIFYING STANDARD FOR COMPOSITION

By 1934, those gathering for the annual Conference on English, according to Chairman Sada Harbarger, considered "a qualifying standard for composition" due to the nature of English after college "in its functional uses" (Report of the Conference on English, 1934, p. 244). This move to a qualifying standard was an important point because English faculty were attempting to sort out the English requirements and their place in the engineering curriculum. Should composition be the sole writing requirement? Did students still require work in literature? Should *technical writing* or *technical exposition* (as it was sometimes called) be taught as a separate course as some universities presently offered it? Even though literature was still being taught, its prominence in the curriculum continued to diminish. In fact, when the members of the SPEE English Committee gathered to talk about English concerns, well over half the time the topic of conversation (and papers) was writing.

The discussion, however, for some years remained stalled at the composition level. Engineering and English educators wrestled with the continued poor performance and lack of interest by their students in composition, with some even describing the course as yet another failure of the high schools to teach the basics. This is not a new argument. Even today educators decry the poor preparation of first-year students for English composition; calls for the elimination of freshmen composition still appear in current journals. In the mid-1930s, however, a potential solution for the engineering English problem was proposed. If only one writing class could be delayed until the junior year of a student's curriculum, why could not that class be specially devised for the needs of the engineering student? Why could not composition provide all students' first-year instruction in writing, but *technical writing* be reserved for later in an engineer's curriculum? In fact, not only did educators see the relevance in such a plan, but increasingly SPEE papers on writing began to reflect the word *technical* in titles.

Thus, postponement, a plan that seemed initially another minor attempt at solving a major problem, indirectly contributed to the shift to a technical writing course, particularly at the upper levels. Papers, in the SPEE and elsewhere, began appearing more regularly with the words *technical writing* or *technical exposition* in the title. Guidelines for the courses emerged, both in George A. Stetson's (1932) "The Art of Technical Writing," and Carl Naether's (1935) "Teaching the Course in Technical Writing." Stetson offered the following condensed "Summary" of technical writing:

1. Become thoroughly familiar with the subject and assemble all the material at your disposal. Do not neglect the literature.
2. Decide upon the purpose and scope of the paper as affected by the occasion and the audience.
3. List the major items to be covered and choose the material to be used.
4. Get a good title, not too long, not too general, but precise and significant.
5. Let the reader know at once what you propose to write about and of any major conclusions that can be effectively divulged.
6. Get his interest at the start and keep it.
7. Have him constantly in mind while you write and do not "write down" to him.
8. Choose some logical order of presentation and stick to it, building on the reader's knowledge and interest.
9. Be careful in the choice in words and in the use of examples and analogies.
10. Be brief, but clear and coherent.
11. Leave out irrelevant details and qualifying phrases that are obviously unessential. Let curves, sketches, illustrations, and tables tell as many of the tiresome details as possible.
12. Give the reader a good bibliography but do not annoy him with too many footnotes.
13. When the paper is written, check it over to see if it is properly proportioned. Pay particular attention to first and concluding paragraphs.
14. Lay the paper aside and read it over several days later to make sure it is complete, well-proportioned, logical and coherent. (pp. 497–498)

Stetson's guidelines, some of which are relevant today, provided, in general, course content for technical writing courses offered to sophomore or junior engineering students.

Alhough Naether (1935) provided similar guidelines for the course, he covered another matter slowly gaining ground among both engineering and engineering English educators. Naether advocated the course in technical writing as a way to segregate engineering students, "since usually the intelligence level of the technical student is much higher than that of, shall I say, the common run of liberal arts students" (p. 650). Just as postponing the writing course turned out to be a factor in the evolution of technical writing as a discipline, so too did segregating the course factor in the eventual move of technical writing as a discipline distinct from composition.

SEGREGATING TECHNICAL ENGLISH

The desire to remove technical writing from the composition requirement grew in popularity as the decade evolved and with good reason. First, removing technical writing and making it a separate, required engineering course made the "qualifying standard for composition" easier to establish because the composition course could return to its original design, a course taught in roughly the same manner at all institutions—an introduction to expository writing. Second, segregating technical writing from composition permitted engineering educators to have a greater role in the development of this still troublesome requirement because the course would closely mirror the engineering curriculum and, indeed, might even be taught as an engineering subject. Third, removing the technical writing course from the composition curriculum meant a course designed specially for those taking degrees in engineering. Thus, all students would be required to take the composition course as freshmen, but the technical writing course could evolve as a specialty for juniors or seniors, who, as we have seen during the discussion of delaying the writing requirement, were much more willing to undertake a semester of writing reports and letters. Such a move seemed nearly inevitable. Engineering educators remained puzzled by the lack of success in English by their graduates. After decades of studying the problem and proposing solutions, the answer, ultimately, would not only change the way engineering students regarded the English requirement, but also signaled the emergence of a new class—technical writing. Could the *field* of technical writing be far behind?

In 1938, Roy J. Barber, writing a piece for the SPEE's new "English Notes"[2] signaled one of the first moves toward segregation of technical writing when he wrote "that technical English can best be taught within the School as an engineering subject, [since] the students' interest may there be aroused by a teacher who is in sympathy with his practical ambitions" (p. 169). The implications and reasoning behind Barber's call are varied. By suggesting that technical English become an engineering subject, Barber (and others who concurred with his idea) implied that at least part of the problem with English education for engineers was a direct result of the English department. That is, the English department, as the past few decades have borne out, was often the site of the problem in developing a writing curriculum for engineers because, simply, the faculty emerged from and remained connected to literary criticism and study. Furthermore, to suggest that technical English become an engineering subject meant that educators advocated segregating engineering students from students in other degree programs, implying that segregation would further contribute to the engineering students' ability to learn this important skill—namely, writing. Finally, to put technical English within the engineering department was to empower that engineering department to shape the manner and style of the course. Of course, inherent in such empowerment was the task of

---

[2] "English Notes," which emerged as a regular feature in the SPEE Proceedings in 1936, offered educators an opportunity not only to submit short pieces on English-related concerns, but provided everyone teaching engineering English a means to read about current English curricular concerns.

construction. Would engineering faculty be responsible for developing and teaching this course? Would English faculty be recruited, hired, and promoted as engineering faculty? Who would train the instructors to teach this new class? Would teachers in technical colleges function any differently than teachers in traditional liberal arts universities with engineering schools? These key questions, tied to the emergence of technical writing as a distinct course (and ultimately, discipline), would preoccupy educators throughout the decade and up to the conclusion of World War II.

THE SITE AND MEANS OF TECHNICAL ENGLISH TRAINING

As engineering educators and English faculty gradually moved toward a technical writing curriculum for engineering students, one problem remained. Where would young, eager English faculty acquire the *technical* training to teach this new course? Such a question, of course, presupposed the distinct disciplinarity of technical writing. This was no longer to be an aspect of composition—a mere 1.40 units of a nearly 10-unit program (according to Fountain). Technical writing was becoming something larger, more than a component of a class, more even than just a class. Technical writing was becoming a field, and the benefits were obvious.

For decades, engineering English educators desired, and in fact needed, more equity and respect from both their English and engineering colleagues. Perhaps this could be accomplished, suggested J. Raymond Derby (1938), Iowa State College professor, in his address to the SPEE on "Improving the Status of English Instructors in Technical Colleges," if the technical writing teacher were properly trained. No such programs at the Master's or PhD level existed for training in a field that was literally just emerging. Derby, however, made a very logical suggestion. The training, he wrote, should occur at

land-grant colleges that stress science and technology but already possess strong departments of English competent to offer majors and the Master's degree. In such surroundings the most alert and experienced professors are likely to see the problem and to be concerned about it (p. 253).

Perhaps indirectly, Derby's suggestions opened two very promising doors for the evolution of this discipline. First, because many land-grant colleges were already experimenting with technical writing, they were the ideal milieu for training a new kind of English faculty member. More important, the faculty who might train technical writing teachers were likely be those who for decades had promoted this kind of education—the trailblazers of a new discipline.

In fact, as Connors (1982) points out in his history of technical writing, the discipline was becoming "a thriving industry in 1938, having produced its own authors, experts, and directors" (p. 338). Formal training at technological colleges seemed likely to benefit engineering education, even while technical communication degrees would at last legitimize those who had been teaching the course in one form or another for years. Their expertise, suggestions, classroom strategies, and perspectives on the unique role of the English teacher in an engineering environment would prove invaluable. Two problems, then, could be solved. At last, educators could look to an engineering English course devised for an increasingly technologically-oriented society. Just as importantly, engineering English faculty might finally receive the respect they had lacked and might come to embrace this discipline to which some of them had been relegated.

Not surprisingly, some resistance to this plan arose. Following Derby's (1938) call for teacher training in technical writing, the president of the SPEE, Karl T. Compton, introduced a piece by Dean H.P. Hammond (1938), in which Hammond noted his "disclaimer that it is the function of the department of English to train students in the writing of technical reports. That, it seems clearly, is the duty of each engineering department" (p. 552). Whether the SPEE president formally endorsed Hammond's idea seems less important than the idea's prominence; "The Function of the Department of English in the School of Engineering" (1938) was the lead article in a report prepared with the assistance of the Committee on English. Hammond went on to note that some English departments offered courses in technical writing that *supplemented* the report writing work of engineering departments. This is interesting because Connors (1982) noted that Fountain's work on technical writing indicated "every technical writing course…covered the report form" (p. 338). None of this is really surprising. Engineers may have felt a proprietary interest in report writing in general. English faculty considering a technical writing curriculum undoubtedly were still determining what could be appropriated from both the composition and engineering disciplines in order to structure this new course and evolving field.

Ironically, the nature of technical writing lent itself to resistance, particularly by the literary set. Technical writing was evolving essentially in response to the engineer's need for functional, practical, real-world writing. Although composition might be described as *functional* writing, its function is to permit students to communicate within the context of a university environment. Technical writing, on the other hand, clearly was a means for students to learn not so much the language of the university community, but to learn how to communicate after graduation. In that way, technical writing's evolution shares a great deal with the evolution of its likely parent discipline—engineering. It must be remembered that engineering was the field perceived by those in both scientific and humanistic pursuits as *vocational* from the mid-19th century onward. Many who teach the subject today know that technical writing shares that same occasional designation of *vocational*, particularly in an environment where literature is sometimes deemed the *prize* in course offerings.

Somehow, there persisted the notion, particularly by those teaching English, that the subjective was ultimately superior to the objective, that function negated creativity, that the practical diminished the humanistic. Indeed, wrote W. O. Sypherd (1939), English professor at the University of Delaware, in his "Thirty Years of Teaching English to Engineers," "we [educators] have reached a sort of impasse in this business of English for the Engineer" (p. 162). The impasse, as Sypherd explained, himself a technical writing textbook author,[3] was related to the problem "of securing proper teachers for a proper English curriculum in an engineering college" (p. 165). As all the pieces gradually came together to shape technical writing as a distinct discipline, one piece remained elusive—the teacher. Educators could settle on aspects of the course in technical writing, and could even settle finally on the need for the course, particularly later in a student's academic career. What no one could settle on by the end of the decade was whether technical writing teachers required special training, and if so, who should offer it.

ENGINEERING ENGLISH IN TECHNICAL VS. LIBERAL ARTS UNIVERSITIES

As the 1930s gradually closed, papers presented at SPEE meetings on the role of the technical writing teacher indicated a desire to establish training guidelines, but no real sense from where those guidelines should emerge. H.L. Creek (1939), in his important piece, "Teachers of English in Engineering Colleges: Selection and Training,"[4] tried to initiate some guidelines for teacher training, after first noting the essential problem: "Although some persons believe that the training in English for engineering students is successful, at many institutions there is enough dissatisfaction with the teachers of English to disturb complacency" (p. 300).[5]

Creek, English Department Head at Purdue, after exploring the dissatisfaction and the existing system of teacher training (nearly exclusively graduate work in literature), and again demonstrating technical writing's ties to composition, argued for "as thorough training in composition as in literature" (p. 308). Certainly composition's own struggle to exist and garner respect in the world of literary study is well documented[6]; as early as 1939, that engineering educators were calling for advanced degree work in composition is encouraging.

The line, however, between these distinct disciplines—composition and technical writing—remained blurred at this juncture in the decade. In fact, Creek concluded by noting that if specialization "is needed, it is in some advanced composition courses that

---

[3] Sypherd, along with Sharon Brown and A. M. Fountain wrote *The Engineer's Manual of English* in the 1930s. The book, which was originally published in 1933 and revised and republished in 1943, guided students through not only letter and report writing, but provided many "specimens" to which students could refer, as well as a special section on writing technical articles for journals.

[4] This selection was Chapter V of the *Report on Instruction in English in Engineering Colleges*, printed in both the SPEE proceedings and consecutive issues of the *The Journal of Engineering Education*.

[5] In fact, Creek (1939) cited engineering students who felt their English teachers were not "masculine" enough, noting in particular one engineer who believed that "men trained in the graduate schools are most certainly unsatisfactory." An English teacher, concluded Creek, cannot simply glide "into a classroom and greet a class of engineers with a sweet schoolgirl smile and 'My, is not this a beautiful spring morning?'" (p. 301).

[6] See both James A Berlin's *Writing Instruction in Nineteenth-Century American Colleges* (1984) and *Rhetoric and Reality: Writing Instruction in American Colleges, 1900–1985* (1987). See also Robert Connors (1986) "Textbooks and the Evolution of the Discipline."

make special applications of English in the writing of technical articles or reports" (p. 308). Technical writing, in the minds of many, continued to exist at best as a specialized form of composition. Something else had to occur to cause these educators to perceive this kind of writing differently, to reconsider the standard expository format for engineers. Creek settled on a phenomenon that would make a difference not only in the way teachers might be trained to teach, but more importantly where they might be trained and how the *location* of this training would influence the move to technical communication in American engineering programs.

Why is it, Creek (1939) wondered, "that the complaints about instruction in English occur more frequently in the engineering colleges of large universities than in separate engineering schools or land-grant colleges, where technology is paramount?" (p. 304). Certainly the distinctions between these two established types of institutions—one which emphasized a variety of educational opportunities and the other which stressed technical education—would not on the surface seem great enough to play a role in the evolution of technical writing as a discipline. However, the differences inherent in the instructional mission of these institutions do lend themselves to this discussion. Simply put, technical colleges and universities had fewer complaints about English instruction than did those institutions in which engineering constituted one of many departments or schools.[7] Perhaps this occurred because technological institutions favored an across-the-board curriculum that emphasized function and real-world context. Perhaps, too, liberal arts institutions believed that English courses devised especially for engineering students was a kind of *favoritism*, even a means of mollifying the vocational students who did not excel in traditional English courses. Perhaps this persistent distaste for any *academic* subject remotely vocational or functional precluded the evolution of a humanistic discipline that would merge technology and writing. Perhaps the question of why one institution had more problems with English instruction than other types of institutions indirectly provided the answer to another, more important question: Where and how should we educate English teachers in this new, evolving discipline called technical writing?

THE ROLE OF SEGREGATION IN
TECHNICAL WRITING'S EVOLUTION

The answer to that question may have been answered by W. Otto Birk (1939), in his piece "Organization and Conditions,"[8] which appeared in the 1939 "English Notes" column. Birk, Professor of English at the University of Colorado, found that "the chief difficulties in all aspects of teaching English to engineering students are in universities with arts colleges" (p. 408). This may have been due, he suggested, to the degree to which engineering students were segregated from other students and the number of

---

[7] See both Fountain's (1938) *A Study of Courses in Technical Writing* and W. Otto Birk's (1939) "Organizations and Conditions."

[8] This piece was Chapter VI of the *Report on Instruction in English in Engineering Colleges.*

English instructors who taught only students in an engineering curriculum. To determine the degree to which segregation of both students and faculty in any way influenced the relative degree of satisfaction with English instruction, Birk sent a questionnaire to 69 colleges and universities.

Sixty-six respondents returned the questionnaire, a high return rate that seemed to confirm the relative importance of the issue. After reviewing the material, Birk established six organizational schemes for offering English instruction:

1. Department is wholly within the College of Engineering, independent of the general or Arts College department.
2. An autonomous staff, nominally a part of the general or Arts College staff, teaches engineering students only.
3. Sections especially provided for engineering students are taught by instructors from the general or the Arts College staff who are deliberately assigned to these sections and who teach engineering students only.
    x. They plan their courses with the wishes of this faculty in mind.
    y. They plan their courses according to the wishes of the general department.
4. Sections especially provided for engineering students are taught by instructors from the general or Arts College staff who are deliberately assigned to these sections, but teach other students also.
5. Sections especially provided for engineering students are taught by instructors from the general Arts College staff who are not assigned. (Sections are given to any instructor who is available.)
6. No method of segregating students is provided. (Engineering students take the course in English provided for other students in the University). (p. 410)

Of the 66 responses, the largest number (29) indicated that, according to Birk, "students are placed indiscriminately with all other students and are taught by members of the department of English" (p. 411). Two points, admittedly unrelated to Birk's goal, but nonetheless relevant to the move to technical writing, arise from this figure. First, the indiscriminate placement of engineering students in with liberal arts students pinpointed the endurance of the lower level composition course plus upper level literature course in many engineering curricula. These engineering students faced the same English courses their nontechnologically oriented colleagues faced. Second, this organizational design for English completely precluded the evolution of technical writing because the need for such a course simply did not exist. Liberal arts students certainly did not require or likely desire such a course aimed at the specialized writing needs of the engineer. Thus, change in English offerings for both engineering students and the eventual instructors of technical writing primarily occurred in the technical universities and land-grant colleges, which, for the most part, segregated both engineering students and the English faculty who taught them.

Naturally, some of Birk's respondents voiced opposition to the notion of segregating students and faculty, although nearly all were from A.B. institutions. Birk noted no such opposition from teachers in technological and land-grant colleges; in fact, he interpreted from his findings "that teachers especially chosen to instruct engineering students or engineering and other students of similar interests are at least likely to handle their students with sympathetic understanding" (pp. 418–419). Birk's point here—that segregated students and teachers each benefited from a streamlined, sympathetic curriculum—is commentary on the willingness of some institutions to develop overall programs that assisted students dealing with historically troublesome classes like English. In addition, his point spoke to the relative satisfaction of both teacher and administrator when segregation occurred.

Birk also discovered that segregation of technical writing faculty impacted promotion from the junior faculty ranks. Consider, for example, the second part of Birk's questionnaire in which he looked at professorial rank for composition and/or technical writing faculty in A.B. universities. Overwhelmingly, most of the teaching was done by assistants and instructors. In fact, "in seven of these schools there is no one above the rank of assistant professor; in two others there is no one above the rank of instructor; and in two others there is no one between a single professor and the instructors" (pp. 427–428). Compounding this problem was the issue of segregation in A.B. colleges. When, in fact, engineering students and English faculty were segregated, the English faculty were overwhelmingly young and inexperienced, virtually stranded at the junior ranks.

Birk's study clearly confirmed several points related to the gradual evolution of technical writing in the 1930s. First, the belief that composition and literature were adequate English requirements for engineering students, particularly in liberal arts universities, undoubtedly slowed the move to any English alternative designed to accommodate the unique needs of the engineer after graduation. Second, experimentation in technical writing curriculums at technological or land-grant institutions provided the impetus to change. Third, the general inequality of English faculty teaching service courses, regardless of institutional mission, slowed the growth of technical writing curricula because no rewards existed for such faculty in either promotion or salary. Finally, because experiments in alternative technical writing curriculums occurred in technological or land-grant colleges, those institutions provided the logical environment for training faculty to teach technical writing. The need for technical writing courses during this decade may indicate movement in the evolution of this type of engineering English offering, but the call, by the end of the decade, for training faculty to *teach* technical writing clearly signals the development of a new discipline.

1940: ENGINEERING ENGLISH POISED FOR CHANGE

By the end of the 1930s, America braced for the inevitable changes that World War II would bring. Change was in store for the engineering community as well. Educators had not yet settled on the best means for solving the still perplexing problem of a coherent and manageable English curriculum for engineering students, but change was imminent. Continued calls were issued by educators for a new kind of English curriculum, even while manufacturers and industrialists demanded people be able to write about America's burgeoning technology.

As Robert Connors (1982) noted, "the first five years of the new decade brought activities that would result in a complete restructuring of engineering education and the beginnings of the final transformation of technical writing courses into the courses we still teach today" (p. 339). Connors believed that the "technological imperative" of World War II contributed to the changes that would occur during the 1940s, a belief shared by Katherine H. Adams (1993), who, in her book *History of Professional Writing Instruction in American Colleges*, noted that "a boost for all types of professional writing instruction occurred [due to] World War II" (p. 149). Both engineering and English educators were poised for change by the end of the 1930s, but how would this change manifest itself? There were suggestions, particularly in the SPEE Proceedings for 1940, that hinted at what was to come.

Universities were gradually moving toward teacher training in technical writing, with industry not far behind. If, as early decade calls indicate, postponement of the engineering English requirement (presumably technical writing) met with more success for both students and instructors due to the maturity of the older, seasoned student, how much more sense would it make, then, for industry to offer postgraduate work in technical writing in-house? In the first Report of the Conference on English for 1940, just such a proposal was introduced. Devised by Westinghouse in conjunction with the University of Pittsburgh, graduate students were able to apply the real-world context of industry writing to their academic advancement. In fact, the University of Pittsburgh offered graduate credit for the course, which placed "emphasis on the shorter, or letter type of report" (p. 303).

The course, described in detail in a separate piece by W. George Crouch and M.L. Manning (1940) entitled "Is Your English Short-Circuited?" reflected, as Connors suggested earlier, elements of technical writing recognizable to those of us who teach the course today. A summary of the piece, devised by the authors as a kind of curriculum outline for educators, is included here in Figure 5.1 (pp. 708–709).

The 1940s, with both the upheaval and challenges that inevitably followed, was nevertheless a decade that witnessed some reconciliation and satisfaction for engineering educators and engineering English faculty. As the next chapter demonstrates, during the 1940s, curricular matters such as the manner and means of teaching English courses in engineering institutions were largely resolved. More importantly, the course taught under so many disparate titles (engineering English, technical composition, technical exposition, technical report writing) at last emerged as the course and discipline we know it to be today—technical writing. Evolving from the early calls for a humanistic means to provide engineers culture, into an element of engineering composition courses, it ultimately appeared as a hybrid course incorporating both the humanistic and functional. Technical writing, more importantly, became not just a course for engineers, but finally a discipline attractive to anyone wishing to teach or write in a technological age.

Circuit Analysis			Protective Methods Used to Minimize Short Circuits
Type of Circuit	Time Allotted (Weeks)	Component Parts of Circuit	
1. Business Correspondence	6	1. Types of Letters Instruction Letters Inquiry Letters Transmittal Letters Sales Letters Interworks Letters District Office Letters	1. Use of actual correspondence letters 2. Analysis of letters for layout, plan, purpose and expression. 3. Style of letter—block, semi-block, indented, military. 4. Study of trite expressions. 5. Emphasis on individuality of expression.
		2. Dictation	1. Practical application of the principles of letter writing. 2. Students assigned letters written in their departments. 3. Replies dictated in class. 4. Specific criticism of letters placed on blackboard. Remedies for each error suggested.
2. The Engineering Report	6	Kinds of Reports 1. Progress 2. Periodic 3. Examination 4. Research	1. Concrete distinction between kinds. 2. Use of standards published by the A.T.E.E. and by the Westinghouse Company. 3. Efficient note taking. 4. Analysis and interpretation of data gathered. 5. Recording of bibliographical items. 6. Use of diagrams, graphs and charts.
3. The writing and the presentation of technical papers	6	1. Preparation and presentation of papers at A.I.E.E. meetings and other technical groups. 2. Engineer obligated to publish results of work. 3. Majority of engineers rely upon published papers for stimulation. 4. Budget limitation, a major item for the publication of technical papers.	1. Experimental papers written in class. 2. Subjects chosen related to interests of class. 3. Particular attention to standards set by A.I.E.E. 4. Brevity with completeness emphasized.
4. Vocabulary	4	1. Insufficient attention to reading and use of the dictionary 2. Lack of an adequate choice of words.	1. Study spread out over year—not localized by concentrated 4 weeks of study. 2. Reading of specific essays. 3. Test for new words—use of dictionary, etymology. 4. Recognition of key words—mental pictures. 5. Stress upon social responsibility

5. Public Speaking	8	1. Failure to organize ideas quickly and effectively.
		2. Explanation of technical subjects in non-technical language
		3. Inability to speak without embarrassment and fluently.
		4. Discussion of the papers presented before Societies.
			1. Combination of speech with written work.
			2. Adaption of material to the audience.
			  (a) How to gain attention.
			  (b) How to talk.
			3. Organization of a talk.
			  (a) Appeal to audience by graphs, charts, and various devices.
			  (b) Analysis into main parts.
			4. Conversational style of delivery.
			5. Demonstration of the principles of public speaking by use of films.

FIGURE 5.1. CURRICULUM OUTLINE FOR EDUCATORS, 1940

Chapter 6

1941–1950:
The Emergence of a
Technical Writing
Discipline

If the 1930s hinted at the full development of technical writing as a discipline, certainly trends emerging from material available in the 1940s indicated some realization of
the discipline's distinct nature and academic viability. For example, by the end of the
1930s, industry entered the discussion of technical writing as a discipline by offering
classes in such writing in house. Connors (1982) suggested that "radical upheaval was
certainly on the way" (p. 339), undoubtedly due to the demands of World War II and
the societal turbulence it would bring. America, by the 1940s, was on the cusp of
change—change brought about by war, change manifested in burgeoning technology,
and change necessitated by an evolving postwar student body at colleges and universities throughout the country. Although the decade was not a period during which scholars and educators produced a great deal of published material on the emergence of
technical writing,[1] articles and texts do reveal momentum toward the emergence of this
distinct discipline. Two very different trends—defense-related writing and the move to
a "humanistic-social" stem in engineering education—provided important linkages in
technical writing's distinctive academic place.

Before analyzing these trends, the impact of World War II on American engineering
education must be considered. The war brought with it not only societal turbulence and
uncertainty, but also advances in technology as manufacturers developed bigger and
more powerful weapons. This meant not only more jobs in manufacturing, but, accord-

---

[1] Connors (1982), in fact, noted in his research that "at least as it appeared in journal articles, [the
engineering English industry came] to an almost dead stop" (p. 339). Although I would not go quite this
far, I also discovered that the rate of articles related to technical writing dropped off, as did the SPEE's
attention to the topic.

ing to Connors (1982), "necessity had mothered thousands of frightful and complex machines, and the need for technical communication had never been greater" (p. 340). Of course, the irony here—that weapons of destruction increased societal prosperity through demands for manufacturing and communications-related positions—could not be overlooked. Papers and reports in the SPEE Proceedings from 1941 to the end of the war reflected some confusion about what the war meant to educators. For example, worries about postwar veteran students appeared, as well as concerns about propaganda—generated both inside and outside of America.[2]

Most importantly, though, the 1940s and World War II brought a technology so advanced, with production so accelerated, that engineers alone were incapable of explaining the changes to the general public. In fact, as material from this decade illustrates, engineers were ill-suited to this business of explaining technology to the public for a variety of reasons, including their close proximity to the product and their tendency to explain at a high level of technicality. As David R. Russell (1991) noted in his *Writing in the Academic Disciplines, 1870–1990,* "technological advances during and after the war created a need for technical writers" (pp. 249–250).

WRITING AND DEFENSE

Defense-related production definitely influenced the development of a technical writing discipline for two reasons. First, as the sophistication of weaponry increased, manufacturers needed writers to explain that technology to personnel who may or may not have had a technical background. Second, engineers, who had previously been largely responsible for writing user documentation to accompany their creations, had only a few English courses to draw on for the challenge of explaining technology to the sometimes technologically ignorant. Thus, the engineers' expertise in devising and creating new products did not necessarily mean that such expertise easily translated into writing. In fact, noted Russell, "corporations and governmental bodies began to hire technical-writing specialists to produce manuals, instructions, reports, proposals, and myriad other documents" (p. 250). Technical writing, then, was realizing full status as a discipline and field because people were being hired to do it; "it was a job in itself," noted Connors (1982, p. 341). The SPEE was also aware of this trend, and in the 1941 Report of the Conference on English noted (in a paraphrase of Jay Gould's "Written English in Defense") "the newly created committee on civilian defense" (p. 278). This committee's duties included: (a) training against subversive propaganda; (b) reports on

---

[2] In one example, Connors noted Fountain describing "the frantic era of the GI Bill, the quonset hut, the barracks classroom, and tar paper apartment, infested by returning veterans" (p. 341). In another example, in Jay R. Gould's (1941) "Written English in Defense," the public is asked to awaken "to a very real danger: the spreading information through harmless gossip. I will not attempt to speak of the problems of official censorship. However, one of the aims of the defense office will be to make each citizen a censor in himself" (p. 527).

activities of groups in patriotic service; (c) using local radio stations for giving out information, instructions, and encouragement.

Jay Gould (1941), who cited this committee in his piece "Written English in Defense," noted also that "every one of these services is based upon the ability to write and speak good English" (p. 523). In addition, then, to explaining technology to the masses, English was also being invoked as a means to educate and encourage society.

Clearly World War II contributed a great deal to education; the conflict, in many ways, provided the impetus for change specifically where technical writing was concerned. This was ironic for a variety of reasons, but most especially because technical writing emerged and evolved, in the years prior to World War II, in a milieu of conflict—the conflict of English and the Humanities in an engineering curriculum, the internal conflicts engineering itself experienced, and the constant conflict of engineering students resisting an English education they considered largely useless. Certainly those advocating a technical writing curriculum, up to this point, fought for both respect and recognition. That the battle abroad made clear to industry the need for this distinct discipline confirmed what many technical writing teachers had previously believed. By synergistically linking engineering principles with a composition curriculum, by devising a way to write technically for a diverse audience, technical writing would no longer be an aspect of another discipline—technical writing would be a discipline.

Perhaps no one better explained the connection between technical writing and defense-related issues than did Jay R. Gould (1941) in his "Written English in Defense." His article even opened with an anecdote in which a friend commented that teachers of English cannot offer much to the national defense. On the contrary, noted Gould:

> Faulty communication, both written and oral, as they affected the army, were very real. Faulty writing means misinterpretations, and misinterpretations result in a lack of efficiency; and efficiency in all its forms is vital to a prepared army. (p. 523)

In addition to benefitting a prepared army, technical writing could also claim the vital nature of efficiency in writing, as well as economy, also alluded to by Gould when he argued that "verbiage should be abhorred, and brevity and coherency stressed as of the most vital importance" (p. 524).[3] Gould stressed these factors, too, because of the need for people to write functional documents—manuals, handbooks, and other user materials for the armed services.

WRITING ABOUT TECHNOLOGY

Gould's belief that the growing need for functional documentation was central to the

---

[3] In fact, so concerned was Gould with the issues of efficiency and economy that he cited Winston Churchill, who, after becoming prime minister, issued "his famous order to the civil service calling for simplification of official language; asking, for example, a substitution of a simple 'Yes' for the official 'The answer is in the affirmative'...The Manchester *Guardian* christened him 'Jack the Jargon Killer' (pp. 523–524).

role of English and/or writing in terms of defense implicitly supported the evolution of technical writing into a distinct field. Consider, for example, Gould's (1941) assertion that "manuals of directions must be geared for...new members of the army." "It is not," he admonished, "as easy as it sounds" (p. 524). Indeed, he was correct. Writing materials for an inexperienced or uninformed group, like new recruits, required not necessarily the expertise of an engineer, but the abilities of a person trained to write about technology according to specific audience considerations.

In 1943, for example, Alexander M. Buchan presented to the SPEE his "English in Basic Army Courses," a paper that dealt not only with the difficulties inherent in teaching new recruits English, but also clearly distinguished between writing about science and technology for a highly technical audience and writing for the public, with a generally low level of technical background. He wrote:

> An important distinction, that is to say, must be made between the mathematical and chemical formulae underlying a technical course and the description of their embodiment in apparatus or in objects of practical use. The aerodynamic theory, for instance, underlying the construction of a propeller, is...quite intricate; but the propeller itself can be described in words—words that may not satisfy a designer but which enable ordinary people to distinguish between a propeller and an egg-beater. (pp. 538–539).

Clearly, Buchan saw the value of writing for the general population in terms that supported the presence of technical writing in an engineering curriculum. To devise a useful object might be the goal of the engineer; to explain the function of that object to its potential user was the realm of the technical writer.

Gould and Buchan succinctly detailed technical writing's emergence in the milieu of war; writing about technology, educators began to determine, was a different enterprise than inventing or developing technology. Thus, gradually, technical writing became not only the means for the engineering student to better grasp and use English writing skills, but technical writing became a means to an end; the discipline superseded its initial goal of contributing to the full education of an engineer. By mid-decade, in fact, reports from two meetings of the 1944 English Conference revealed not only the close connection of English and engineering,[4] but the secretary of the 1944 Conference on English, Karl O. Thompson, derived the following list of concerns from the various papers presented. The list clearly acknowledged the role of *technical* writing, while also hinting at other shifts in English education during the latter half of the decade:

1. Practice in student reading aloud is helpful and must be made permanent.
2. Reading as a basis for careful speaking and writing is neglected.
3. The concentrated plan of study has proved to be effective.
4. Report writing and other technical principles cannot be taught to freshmen.

---

[4] Phillip W. Swain, editor of *Power Magazine*, for example, delivered a paper called "English, the No. 1 Vocational Study for Engineers."

5. Courses in English need to be given in each of the four college years.
6. The variety of student backgrounds in the same classrooms makes teaching difficult.
7. Adaptation of English to the engineer must be made.
8. Some found the attitudes of service men were not studious or conducive to excellence of work.
9. Some good points of army and navy programs should be continued.
10. Returning soldiers can be helped by courses in literature.
11. Recognition of special abilities among students must be increased.
12. Students can learn fast if they must.
13. The close relation between logical thinking and clear writing must be stressed.
14. There is danger in substituting courses in history, government, and geography for courses in English.
15. Definite standards in English for graduation must be maintained. (pp. 94–95)

This list, which detailed concerns related to technical writing, the role of the soldier-student, and the role of literary study, preoccupied the SPEE until the end of the decade. This list also suggested a belief that English education needed to be standardized to include specific writing elements, partially motivated by the return to school of war veterans with little academic training. Technical writing educators also were looking more closely at the standard elements of their emerging field. Papers presented to the SPEE membership during the last six years of the decade revealed technical writing trends that still exist in the curriculum today.

STANDARD PRACTICES IN TECHNICAL WRITING

The gradual evolution of any discipline must necessarily bring with it the gradual emergence of discipline-specific features. Technical writing certainly was no exception. Although publication of textbooks continued to flourish during the 1940s,[5] those texts essentially addressed the same issues as previous texts had covered—namely, format styles for report, letter, and article writing. Papers presented on technical writing to the SPEE during the mid- to latter portion of the decade revealed a degree of specificity in teaching technical writing that had not existed before.

Previously authors wrote of audience concerns and report styles, but with Joseph

---

[5] Nelson's *Writing the Technical Report* appeared in 1940, as did Oliver's *Technical Exposition*. These two texts were largely in keeping with the books published thus far, with chapters on report writing, business letter style, and variations of the technical description and definition papers students construct today. Kapp's *The Presentation of Technical Information* was first published in 1948. This book, different from the others published in the 1940s, was based on four lectures given at the University of London, where Kapp was Dean of the Faculty of Engineering. Kapp was quite an advocate of functional language, arguing that "it would be a mistake to suppose that, in order to train a student in the use of Functional English, it suffices to make him acquainted with the best in English literature" (p. 9).

Jones's (1944) "Visual Aids in Technical Writing at the University of Texas," SPEE members encountered a degree of discipline specificity which suggested the complete viability of technical writing in an engineering curriculum, while indicating also that English educators were considering the full ramifications of the emerging discipline. For example, Jones, Assistant Professor of English, described English 317, Technical Writing for Engineers and Science Majors, taught by members of his department at the University of Texas. Jones advocated the use of visuals in both teaching and writing technical documents. The purpose of his piece was to suggest both the usefulness of having students use visuals in their reports and during oral presentations of their reports. This fully acknowledged feature of technical writing today certainly found its roots in the pedagogical practices of teachers like Jones who adapted his curriculum to the still emerging technical writing field.

Similarly, Frederick Abbuhl's (1945) piece, "A Writing Laboratory Course," suggest-

PROBLEM NO. 15

Purposes:
1. To give training in tact.
2. To give training in approaching prominent people.
3. To give training in persuasion.

Assumptions:
1. As secretary of a local society of engineers, you have been asked to invite a prominent engineer to address your group.
2. You do not have funds available for paying a speaker.
3. You do, however, have funds enough to pay travelling expenses.

The Problem:
1. Write the letter inviting the prominent engineer to address your group.
2. Make the engineer understand that his travelling expenses will be paid, but that he will receive no fee for speaking.
3. Make him feel that addressing your group is worth the time and effort he will spend.
4. Be courteous, cordial, and tactful in your letter.

PROBLEM NO. 21

Purposes:
1. To give practice in interesting laymen in a technical fact.
2. To give practice in stimulating the reader's emotions.

Assumptions:
1. That during the year more pedestrians are killed by automobiles in the month of December than in any other month.
2. That most of the deaths occur between the hours of six and eight in the evening.
3. That the article will be published early in December.

The Problem:
Write a 250 word article that will by arousing emotion make the reader careful during the month of December.

Procedure:
1. Devote the first 125 words to dramatization; that is, a single scene such as might appear in a short story.
2. Devote the next 75 words to a statement of the facts given in the assumption.
3. Devote the next 25 words to a direct exhortation to the reader.
4. Devote the last 25 words to a reference to the dramatization and a final warning.
5. Number the four sections.
6. Pick a title which will attract the reader's attention.
7. Count the words and put the total in parentheses at the end of the article.

FIGURE 6.1. CASES FOR BEGINNING WRITERS

ed to SPEE members the usefulness of case studies so that technical writing students could learn to work in the context of the workplace. Abbuhl, Acting Head of the Department of English at Rensselaer, presented several cases for beginning writers, including the two included in Figure 6.1. Although contemporary case studies texts are not as popular as other technical writing texts, such books do exist. Case studies, however, as features of other texts,[6] abound in technical writing programs around the country.

Finally, Herman A. Estrin (1948–49), English Instructor at Newark College of Engineering, presented to the then ASEE[7] his "Occupational Survey as a Topic in English." In the paper Estrin described a sophomore technical writing requirement involving an occupational survey presented in report form. This report on career opportunities for engineers, not unlike similar report types included in introductory technical writing today, urged students to consider factors such as career history, job requirements, qualifications, placement opportunities, demand, and average salary.[8] Although discussions of report types, visuals, or case studies may not seem particularly influential, those papers on specialized, narrowed aspects of a technical writing curriculum may have indicated the discipline's acceptance by the engineering community, as well as its continued relevance in the minds of many educators.

THE PLACE OF TECHNICAL WRITING IN THE HUMANISTIC STEM

This seeming full acceptance of technical writing as the engineering English course, though, was not completely without potential problems. Although English educators continued to write about and discuss the relevance of the course to engineers, as well as technical writing's relevance as a separate discipline, the end of World War II brought with it a kind of backlash against curriculums that emphasized the essentially technical and vocational. Two reports—one in 1940 and one in 1944—suggested that the humanistic was getting lost in the creeping miasma of practicality. The Hammond Reports[9] argued for two stems—one scientific and one humanistic. This argument, largely between English faculty who believed that the humanities were underemphasized and engineering faculty who were primarily concerned with teaching science and technology, would not be adequately resolved. Interestingly, Connors (1982), in his history, pointed out "that neither freshmen composition nor technical writing courses

---

[6] See John M. Lannon's text, *Technical Writing* (1994), written for the beginning technical writing student. Lannon makes extensive use of cases, particularly in exercises included at the end of most chapters. See also *The Case Method in Technical Communication: Theory and Models* (1989), edited by R. John Brockmann and published by the ATTW (Association of Teachers of Technical Writing).

[7] The SPEE (Society for the Promotion of Engineering Education) became the ASEE (American Society for Engineering Education) in 1947.

[8] Interestingly, Estrin (1948–49) asked students to determine the average salary "of men. Of women. Of beginning workers. Of experienced workers" (p. 254).

[9] The reports were directed and overseen by H.P. Hammond and his committee.

were claimed or championed by either side" (p. 340). As a review of the dialogue on the proposed "stems" reveals, although neither the humanistic English teachers nor the engineering faculty wanted to call technical writing their own, they had in the discipline a means for bridging what they undoubtedly perceived as diametrically opposed concerns. Technical writing, as its gradual evolution has demonstrated, offered students the humanistic in a technologically-based framework.

Even a cursory review of engineering education material from roughly middecade through 1950 reveals the preoccupation of English teachers with what they perceived as a diminished role for the arts and humanities in engineering curriculums across the country. Was this a backlash against the very utilitarian education students received during the war years? Was this in response to the ever-decreasing importance of literature in engineering programs? Or was this the concern of English faculty that their roles as teachers in engineering colleges had been gradually reduced to composition (service) courses? Possibly all these factors influenced both faculty and the members of H.P. Hammond's committee when, in 1940, they called for "the parallel development of the scientific-technological and the humanistic-social sequences" (quoted in Connors, 1982, p. 340). The SPEE recognized that the report was "a distinct advancement," noted Karl O. Thompson (1945), because "such action has long been desired by teachers of English in engineering colleges" (p. 163).

English teachers wanted to feel that they were contributing to more than just students' writing; many wanted to feel as well that they contributed to the formation of students' character. Thus, designation and acceptance of a humanistic-social stem was a means to address what was perceived as missing from an engineering curriculum. Ironically, this same issue drove engineering English education during the 19th and early 20th century as educators wrestled with the correct curriculum—one purely vocational or one devised to shape the whole student. By 1940, and after two world wars, English educators in particular wanted to address the humanistic, and so the Hammond Report became a powerful tool for incorporating more literature and history into engineering students' studies. By 1944, when the second Hammond Report was presented, engineering educators actively voiced their displeasure at the emphasis on humanism,[10] but during the postwar academic juggling of programs, "the conception of the humanistic stem won out gradually" (Connors, 1982, p. 340).

Thus, by the end of the 1940s, the old arguments reemerged. Should the engineer pursue a purely professional curriculum? Should English education serve to humanize engineering students? Should engineering students remain in school longer too, if nec-

---

[10] In addition to a spate of articles printed in *The Journal of Engineering Education* during the 1940s, consider the tone of Clement J. Freund's (1943) SPEE presentation entitled "Back to the Professional Degree": "Most urgently, with all the power I can bring to bear, I recommend that the seasoned and experienced leaders in engineering education...give serious thought to the vigorous revival of the professional degree which [they have] either completely discarded or sustain merely in the form of familiar bulletin announcements" (p. 228).

essary, in order take an increased percentage of culture courses?[11] Although none of these questions can be pinpointed as the one, key question, all point to the desire of some educators to synthesize in the engineering student the fullness of technological training and humanizing instincts, including the ability to read and write well. Engineers, wrote James H. Pitman (1948–1949), in his piece on the selection of English faculty for the engineering college, should be "good men in society" (p. 319).

Engineering education questions by 1950, then, began to resemble engineering education questions in the 19th century in that the complete engineer became the concern of many involved in engineering curricular decisions. Technical writing and composition, both accepted and flourishing disciplines,[12] continued to evolve, but neither fit neatly into either the scientific nor humanistic stems.[13] This is ironic because technical writing, in particular, provided the means to do just what some English and engineering educators wanted the humanistic stem to provide—a means to merge in one individual the best of a technologically based education and the best of a humanistic grounding.

THE HUMANISTIC STEM AS TECHNICAL WRITING MATRIX

The desire for a humanistic stem for engineers, I believe, was to prepare them in the Greek ideal of service and productivity to self and society. Whether this desire was fueled by world wars, or whether this desire was rooted in the persistent belief that any higher education must ultimately be tied to the liberal arts, is impossible to say. Consider, though, a text published early in the 1950s, *Engineers as Writers*, by Miller and Saidla (1953). This book did not fit the mold of any earlier technical writing texts because it was written "for the purpose of motivating engineering students by a study of the structure and style employed in reports by men of eminent engineering stature down through the ages" (p. iii). Although structure and style are important to the authors, motivation through historical sources was equally important, thus the social-historical link. Miller and Saidla, in effect, produced a book significantly ahead of its time, a book which demonstrated that not only had technical writing existed historically, but that the writings of engineers, architects, and inventors provided that all important link between the technical and the humanistic. To write about technology

---

[11] Karl O. Thompson (1945) noted, shortly after recognition of the first Hammond Report, that "time allotted to subjects in the Humanistic-Social Division has been from ten to seventeen percent" (p. 163). Connors (1982) noted that by 1944, the second Hammond Report advocated "a four-year program that required 20 percent or more of the student's time to be devoted to humanistic-stem courses" (p. 340).

[12] The next decade, the 1950s, noted Connors (1982), would see the "foundation of the Society of Technical Writers and the establishment in 1958 of the influential Transactions on Engineering Writing and Speech" (p. 342).

[13] In fact, said Edwin S. Burdell (1942), Director of the Cooper Union for the Advancement of Science and Art, during a 1942 roundtable discussion, "I would like to ask a question…regarding the objectives in the humanistic-social stem, and I do not mean by that in English composition, because I would rule that out and classify it with drawing and mathematics" (p. 770).

presupposed the human link; writing that explained a process or that was to be used by the general public reached out to a culture in a way that even literature could not. Technical writing, functional writing, acknowledged society by its very existence.

Interestingly, Miller and Saidla's book opened with a selection from the *De Architectura* by Vitruvius, written circa 27 B.C. The beginning of the selection, "The Education of the Master Builder," clearly expounded on issues relevant in the humanistic-social stem:

> The architect should be equipped with knowledge of many branches of study and varied kinds of learning, for it is by his judgment that all work done by other arts is put to text. This knowledge is the child of practice and theory... It follows, therefore, that architects who have aimed at acquiring manual skill without scholarship have never been able to reach a position of authority to correspond to their pains, while those who relied only upon theories and scholarship were obviously hunting the shadow, not the substance. But those who have a thorough knowledge of both, like men armed at all points, have the sooner attained their object and carried authority with them. (p. 14)

Vitruvius, in effect, described the importance of both theory and practice, of technology and humanism, merged in the individual. Technical writing, initially the answer to the engineering English dilemma, provided the means to this end; that educators did not consider the discipline's broader context, its ability to bridge technology and humanism, was somewhat surprising. That educators did not necessarily perceive technical writing's full potential did not preclude it or composition from being taught in engineering schools; in fact, both were taught regularly,[14] but neither fit neatly into the framework laid out in the two Hammond Reports. Technical writing, now an emerging discipline, should have been the cornerstone of the humanistic stem. That it was not is only one more irony in its evolution as a discipline.

The problem for technical writing was its nature as a discipline. Although it clearly emerged in the engineering schools as a means for students to write about growing technology, the evolving parameters of technical writing expanded to include issues such as audience analysis, reader interpretation difficulties, translation considerations, and rhetorical strategies. Clearly these are not purely technical; these and other features cross over into the rhetorical, the humanistic. Technical writing, by the end of the 1940s, was neither essentially scientific nor completely humanistic. The discipline bridged both and yet was claimed by neither, a phenomenon recognizable even today. This is not to suggest that English and engineering educators did not perceive both the exclusivity of two *stems* and the difficulty of *fitting* disciplines into those stems—disciplines, like technical writing, which did not always neatly fit.

---

[14] Boarts and Hodges (1946), in their *Journal of Engineering Education* article, "The Characteristics of the Humanistic-Social Studies in Engineering Education: A Report," studied the humanistic-social stem at eight well-known private and public colleges. They discovered that not only was composition a first-year requirement at all, but at the Case School of Applied Science, technical exposition was a three-credit requirement.

INTEGRATION: THE MARRIAGE OF SCIENCE
AND HUMANISM

Calls for integration concluded the 1940s, calls which ultimately hint at technical writing's future as a discipline. Pitman, for example, in his 1948 piece called for a "meeting of the minds" and urged educators seeking stems in the curriculum to remember "that an engineering college is just what it purports to be—a place where the *application* of knowledge is all-important." "The Humanities," he continued, are "islands of peace," [but] "they are islands, not the whole continent" (p. 440). Pitman clearly expressed the concern that many had regarding the role of the humanistic in an engineering curriculum, concerns expressed as early as the mid-19th century. Engineering colleges did not need to reinvent themselves according to the standards of the liberal arts colleges; they were academically distinct, even though they overlapped. The problem, of course, was similar to the cooperative plan discussed in Chapter 4. Initially, both cooperation and the two stems presupposed the distinctive and separate nature of disciplines. In fact, the technological and the humanistic need not have remained separate. In the full context of serving to develop a well-prepared engineer, the two stems were inextricably linked.

Burdell (1947), in his "Philosophy of Humanistic-Social Studies in Engineering Education," recognized this when he called for a kind of "scientific humanism [which] means a marriage of science and the accumulated culture of the past; it cannot tolerate separate disciplines moving along parallel lines in accepted grooves of departmentalization" (p. 594). Technical writing, as demonstrated thus far, might have supplied the key to Burdell's marriage of science and humanism. Technical writing, writing about technology, bridged the subjective and the objective, the empirical and the rhetorical.

TECHNICAL WRITING IN POSTWAR AMERICA

As postwar America's technology evolved, however, technical writing's place would be secured. In fact, as Burdell eloquently noted in his 1947 article, "the fabrication and detonation of the atom bomb last summer put the world into a short-sheeted bed" (p. 594). The world had changed, weapons had changed, and technology had changed at a frightening pace. Thus, a technologically based society ought to preclude divisions, particularly between the scientific and the humanistic; disciplines that bridged divisions would be the educational future of colleges and universities, disciplines such as technical writing, as well as contemporary pedagogical shifts in composition, including writing across the curriculum, multiculturalism, discourse communities, and other trends that fully acknowledge that absolute divisions between and among disciplines do not best serve the student.

Technical writing, however, did not evolve into the discipline it is today without periods of division. The discipline, particularly early in its evolution, emphasized the scientific over the humanistic to a degree; logical positivism, now decried as a rigid and narrow means of regarding information transfer, dominated texts and theory as a result. Technical writing grew and evolved in a manner not unlike engineering itself. Through periods of self-doubt, through periods when alternately English or engineering educators doubted the usefulness of the discipline, through periods when any discipline remotely practical was relegated to the simply vocational, technical writing continued to evolve, particularly in the classrooms and the texts of those early teachers such as Earle, Harbarger, Nelson, Rickard, and others who perceived the necessity and unique nature of their undertaking. The divisions of the 1940s may have seemed a particularly precarious time for this still-emerging discipline, but the period, in many ways, only enhanced technical writing's future. Although the humanistic stem placated English teachers in engineering colleges for awhile, Connors (1982) noted that by the middle of the 1950s, "new graduates still could not write well, [so] by 1957, nearly all colleges offered a technical writing course" (p. 342). Technical writing's future was secure.

TECHNICAL WRITING IN A CONTEMPORARY CONTEXT

Even though this study examines the evolution of technical writing from 1850 to 1950, certainly the century of greatest change in the discipline, the nearly 45 years that follow the conclusion of this work, require examination as well. Technical writing's place in English, Humanities, and Engineering departments around the United States needs further study because although the conclusion of this chapter would imply that the discipline's place became secure by the 1950s, this is only partially true. While Connors (1982) wrote that nearly every college was offering the course by the mid-1950s, Souther (1989) noted that only 20 years later (in 1976) the "Society for Technical Communication listed only 19 academic degree programs in technical communication" (p. 2). This raises two important questions for future study. First, at what stage in the evolution of this discipline did technical writing become an academic degree program, and who offered the program initially? Second, at what stage in the move to technical writing did liberal arts colleges, without engineering curriculums in place, appropriate the course for liberal arts majors?

In addition to these questions, and many others that will be answered by scholars in the decades to come, are just a few other points that warrant consideration. Contemporary writer Carolyn Miller (1979), in her "Humanistic Rationale for Technical Writing," noted that the discipline called *technical writing* has been so "shot through" with positivistic assumptions that it is often relegated to the category of a skills course by many English and Humanities departments. She believes instead that "good technical writing becomes, rather than the revelation of absolute reality, a persuasive version of experience" (p. 616). The emphasis on persuasive (rhetorical) elements in her article helped to establish the basis of an ongoing discussion among technical communications scholars—a discussion that requires future research.

Is technical writing essentially a science-based undertaking of simple information transfer, or is such writing, as David Dobrin (1983) calls it, "writing that accommodates technology to the user" (p. 242)? What is to be our definition of technical writing? This study points to technical writing as a discipline with a legacy of both scientific and humanistic stems; it is ultimately a recursive discipline that weds empiricism and rhetorical theory. However, future researchers must analyze the divisions that still exist and often relegate technical writing to a skills course status. If technical writing, as this study has argued, emerged as an engineering composition hybrid, must technical writing, in a contemporary framework, acknowledge composition trends, theory and practice, in order to continue evolving?

In the future, researchers must continue to evaluate the divisions between literature and technical writing (or composition) teachers in English and Humanities departments in this country. As this study has shown, engineering English in the early 20th-century was almost certainly, in part, a kind of literary requirement necessitated by the lack of culture in engineering students. How far have we really evolved when, as Carolyn Miller (1979) noted at the beginning of her "Humanistic Rationale for Technical Writing," "technical writing could not be allowed to serve as a humanities course" [at North Carolina State University in Raleigh] (p. 610)?

From the early "culture" courses of the 19th century, to the move toward practicality and function at the turn of the century, English in engineering colleges sought to provide the engineering student both a sense of social worth and skill in writing. From the 1920s, when pioneers such as Samuel Earle and Sada Harbarger argued for a new type of writing, to the 1930s, when calls for cooperation revolutionized the engineering English classroom, technical writing existed as a possibility not just for engineers, but for any student seeking to bridge the communication gap between inventor and consumer. By mid-century, 1940–50, World War II and advanced weapons of destruction brought to fruition both the need and the means for this new kind of curriculum. With its roots in both engineering and the humanities, technical writing is a unique discipline that exists as an argument against the linearity of academic undertakings, proving instead that as disciplines evolve they must include, not exclude, must bridge rather than separate.

CONCLUSION

This study of the move to technical communication in American engineering programs has explored not only the discipline's roots in engineering composition pedagogy, but has as well detailed the trends, through the decades, that have cumulatively formed technical writing into the discipline we know it to be today. This study has also established the sometimes turbulent history of a discipline which often found itself at the nexus of change—neither composition nor literature, ultimately claimed by neither the humanists nor the scientists. Although technical writing emerged as a reconceptualized composition course, after calls for English and engineering cooperation made traditional composition offerings less relevant, it remained for nearly a century a discipline seeking to define itself. With teachers poorly trained to teach it and textbooks struggling to fill in the gaps, technical writing emerged ultimately as the synergistic result of engineering and English faculty cooperating to create a writing alternative for engineers. Perhaps even inadvertently engineering English became the locus of a new discipline, with academic programs growing up around it.

Although technical writing was born ultimately of need, its potential in a technological society became larger than the need it was originally devised to fulfill. Scholars still have much to uncover regarding technical writing's contemporary presence in English and Humanities departments, not to mention its future in a constantly changing, technological society. Although any variety of factors may have influenced the discipline's evolution, I believe that technical writing's pedagogical roots may exist in the engineering colleges, themselves the site of great changes during the 20th century. Understanding the factors influencing change in engineering curricula reveals the engineering English trends that shaped technical writing. This information must surely play a role in the next hundred years of technical writing's evolution in America.

References

Adams, K. (1983). *A history of professional writing instruction in American colleges: Years of acceptance, growth, and doubt*. Dallas: Southern Methodist University Press.

Abbuhl, F. (1945). A writing laboratory course. *Proceedings of the Society for the Promotion of Engineering Education, 53*, 269–271.

Aldrich, W.S. (1894). Engineering education and the state university. *Proceedings of the Society for the Promotion of Engineering Education, 2*, 268–292.

Allen, J. (1992). Bridge over troubled waters? Connecting research and pedagogy in composition and business/technical communication. *Technical Communication Quarterly, 1*, 5–26.

Alred, G.J., Reep, D.C. & Limaye, M.R. (1981). *Business and technical writing: An annotated bibliography of books, 1880–1980*. Metuchen, NJ and London: Scarecrow.

Aydelotte, F. (1923). *English and engineering*. New York: McGraw-Hill.

Baker, I.O. (1895). Specifications for text-books. *Proceedings of the Society for the Promotion of Engineering Education, 3*, 111–123.

Baker, I.O. (1900). Address by the President. *Proceedings of the Society for the Promotion of Engineering Education, 8*, 11–27

Baker, R.P. (1919). *Engineering education: Essays for English*. New York: Wiley.

Baker, R.P. (1932). Problems of administering English work in engineering colleges. *Proceedings of the Society for the Promotion of Engineering Education, 40*, 282–291.

Baldwin C.S. (1906). Freshmen English II. *Educational Review, 12*, 485–499.

Barber, R.J. (1938). Suggestions for improvement of written English in American schools of engineering. *English Notes. Proceedings of the Society for the Promotion of Engineering Education, 46*, 169–170.

Benjamin. (1908). Discussion. *Society for the Promotion of Engineering Education, 16*, 88–89.

Berlin, J.A. (1987). *Rhetoric and reality: Writing instruction in American colleges, 1900–1985*. Carbondale: Southern Illinois University Press.

Berlin, J.A. (1984). *Writing instruction in nineteenth century American colleges*. Carbondale: Southern Illinois University Press.

Birk, W.O. (1939). Organization and conditions. *English Notes. Proceedings of the Society for the Promotion of Engineering Education, 47*, 408–434.

Bledstein, B.J. (1976). *The culture of professionalism: The middle class and the development of higher education in America*. New York: Norton.

Boarts, R.M., & Hodges, J.C. (1946). The characteristics of the humanistic-social studies in engineering education: A report. *Journal of Engineering Education, 36*, 339–351.

Breitenbach. (1908). Discussion. *Society for the Promotion of Engineering Education, 16,* 89–92.

Brockmann, R.J. (1983). Bibliography of articles on the history of technical writing. *Journal of Technical Writing and Communication, 13,* 155–165.

Brockmann, R.J. (1989). *The case method in technical communication: Theory and models.* St. Paul, MN: Association of Teachers of Technical Writing.

Brown, C.C. (1896). On the desirability of instruction of undergraduates in the ethics of the engineering profession. *Proceedings of the Society for the Promotion of Engineering Education, 4,* 242–249.

Buchan, A.M. (1943). English in basic army courses. *Proceedings of the Society for the Promotion of Engineering Education, 51,* 535–540.

Buck, G. (1901). Recent tendencies in the teaching of college composition. *Educational Review, 22,* 371–383.

Burdell, E.S. (1942). The humanistic-social studies in engineering education, factors and influences at work. *Proceedings of the Society for the Promotion of Engineering Education, 50,* 770–774.

Burdell, E.S. (1947). The philosophy of humanistic-social studies in engineering education. *Journal of Engineering Education, 37,* 593–600.

Burr, W.H. (1921). Some features of engineering education. *Proceedings of the Society for the Promotion of Engineering Education, 29,* 64–79.

Calvert, M.A. (1967). *The mechanical engineer in America, 1830–1910.* Baltimore: Johns Hopkins Press.

Caullery, M. (1922). *Universities and scientific life in the United States.* (J.H. Haughton & E. Russell, Trans.). Cambridge: Harvard University Press.

Chatburn, G.R. (1907). A combined cultural and technical engineering course. *Proceedings of the Society for the Promotion of Engineering Education, 15,* 222–229.

Clark, J.J. (1910). Clearness and accuracy in composition. *Proceedings of the Society for the Promotion of Engineering Education, 18,* 333–342. Discussion. *SPEE, 18,* 342–357.

Committee on English. (1924). *Proceedings of the Society for the Promotion of Engineering Education, 32,* 560–561.

Committee on English. (1925). *Proceedings of the Society for the Promotion of Engineering Education, 33,* 650–652.

Committee on Entrance Requirements (English). (1901). *Proceedings of the Society for the Promotion of Engineering Education, 9,* 278–282.

Conference of Teachers of English. (1923). *Proceedings of the Society for the Promotion of Engineering Education, 31,* 248–252.

Connors, R.J. (1982). The rise of technical writing instruction in America. *Journal of Technical Writing and Communication, 12,* 329–352.

Connors, R.J. (1986). Textbooks and the evolution of the discipline. *College Composition and Communication, 37,* 178–195.

Creek, H.L. (1939). Teachers of English in engineering colleges: Selection and training. *English Notes. Proceedings of the Society for the Promotion of Engineering Education, 47,* 300–313.

Crouch, W.G. (1910). Discussion. *Society for the Promotion of Engineering Education, 18,* 345–346.

Crouch, W.G. & Manning, M.L. (1940). Is your English short-circuited? *Proceedings of the Society for the Promotion of Engineering Education, 48,* 700–711.

Derby, J.R. (1938). Improving the status of English instructors in technical colleges. *English Notes. Proceedings of the Society for the Promotion of Engineering Education, 43,* 252–256.

Diemer, H. (1909). Employers' requirements of technical graduates. *Proceedings of the Society for the Promotion of Engineering Education, 17,* 172–178.

Dobrin, D.N. (1983). What's technical about technical writing? In P. Andersen et al. (Eds.), *New essays in technical and scientific communication: Research, theory, practice.* (pp. 227–250). New York: Baywood.

Drane, W.H. (1910). Discussion. *Society for the Promotion of Engineering Education, 18,* 349–351.

Earle, S.C. (1911). English in the engineering schools at Tufts college. *Proceedings of the Society for the Promotion of Engineering Education, 19,* 33–47.

Earle, S.C. (1911). *The theory and practice of technical writing.* New York: Macmillan.

Eddy, H.T. (1897). Address by the President. *Proceedings of the Society for the Promotion of Engineering Education, 5,* 11–15.

English Department, The. (1925). *Proceedings of the Society for the Promotion of Engineering Education, 33,* 324–332.

English and other Teaching. [Editorial]. (1908). *The Nation, 19,* 253–254.

English in college. [Editorial]. (1915, July 1). *The Nation, 101,* 14–15

Estrin, H.A. (1948–49). An occupational survey as a topic in English. *Proceedings of the American Society for Engineering Education, 56,* 253–254.

Faig, J.T. (1913). The effect of cooperative courses upon instructors. *Proceedings of the Society for the Promotion of Engineering Education, 20,* 97–106.

Fountain, A. M. (1938). *A study of courses in technical writing.* Raleigh: University of North Carolina.

Freund, C.J. (1943). Back to the professional degree. *Proceedings of the Society for the Promotion of Engineering Education, 51,* 223–229.

Frost, H. (1911). Discussion. *Society for the Promotion of Engineering Education, 19,* 54–90.

Gehring, H.A. (1911). Discussion. *Society for the Promotion of Engineering Education, 19,* 54–90.

Gould, J.R. (1941). Written English in defense. *Proceedings of the Society for the Promotion of Engineering Education, 49,* 523–528.

Graham, J.W. & Barton, W.H. Jr., (1931). Practical English course for senior engineers. *Proceedings of the Society for the Promotion of Engineering Education, 39,* 604–606.

Grego, R.C. (1987). Science, late nineteenth-century rhetoric, and the beginnings of technical writing instruction in America. *Journal of Technical Writing and Communication, 17,* 63–78.

Guralnick, S.M. (1979). The American scientist in higher education. In N. Reihgold (Ed.), *The sciences in the American context: New perspectives.* (pp. 99–141). New York: Smithsonian Press.

Hall, A.V. (1931). English as an essential part of the engineering curriculum. *Proceedings of the Society for the Promotion of Engineering Education, 39,* 416–422.

Hammond, H.P. (1938). The function of the department of English in the school of engineering. *English Notes. Proceedings of the Society for the Promotion of Engineering Education, 46,* 550–553.

Hammond, H.P. (1940). Report of Committee on Aims and Scope of Engineering Curricula. *Journal of Engineering Education, 30,* 1940.

Harbarger, S.A. (1920). The qualifications of the teacher of English for engineering students. *Proceedings of the Society for the Promotion of Engineering Education, 28,* 299–306.

Harbarger, S.A. (1928). *English for Engineers* (2nd ed). New York: McGraw-Hill.

Harvard Report. (1893). Report of the Committee on Composition and Rhetoric. Cambridge, MA: Harvard University.

Hofstadter, R. & Smith, W. (Eds.). (1961). *American higher education: A documentary history* Vols. I and II. Chicago: University of Chicago Press.

Hotchkiss, G.B. (1909). *Business English.* New York: New York University Press.

Howell, A.C. (1927). English for engineers at the University of North Carolina. *Proceedings of the Society for the Promotion of Engineering Education, 35,* 621–628.

Hull, D.L. (1987). *Business and technical communication: A bibliography 1975–1985.* Metuchen, NJ: Scarecrow.

Johnson. (1897). Discussion. *Society for the Promotion of Engineering Education, 5,* 229–232.

Johnson, J.B. (1893). Methods of studying current technical literature. *Proceedings of the Society for the Promotion of Engineering Education, 1,* 265–269.

Johnson, J.B. (1894). The teaching of engineering, specifications and the law of contracts to engineering students. *Proceedings of the Society for the Promotion of Engineering Education, 2,* 109–113.

Johnston, T.J. (1903). Engineering English. *Proceedings of the Society for the Promotion of Engineering Education, 11,* 361–371.

Jones, J. (1944). Visual aids in technical writing at the University of Texas. *Proceedings of the Society for the Promotion of Engineering Education, 52,* 292–294.

Kapp, R.O. (1948). *The presentation of technical information.* London: Constable and Co.

Kent, W. (1908). Results of an experiment in teaching freshmen English. *Proceedings of the Society for the Promotion of Engineering Education, 16,* 74–83. "Discussion." *SPEE, 16,* 84–97.

Kent, W. (1916). Discussion. *Society for the Promotion of Engineering Education, 24,* 183–193.

Kidwell. (1897). Discussion. *Society for the Promotion of Engineering Education, 5,* 225–226.

Lannon, J.M. (1994). *Technical writing.* New York: HarperCollins.

MacClintock, P.L. (1914). *The essentials of business English.* Chicago: LaSalle Extension University.

Magruder, W.T. (1916). Discussion. *Society for the Promotion of Engineering Education, 24,* 183–193.

Marvin, F.O. (1894). Common requirements for admission to engineering courses. *Proceedings of the Society for the Promotion of Engineering Education, 2,* 39–58.

Marvin, F.O. (1901). Address by the President. *Proceedings of the Society for the Promotion of Engineering Education, 9,* 13–24.

McDonald, P.B. (1920) Engineering English. *English Journal, 9,* 588–590.

Mendenhall, T.C. (1897). The efficiency of technical as compared with literary training. *Proceedings of the Society for the Promotion of Engineering Education, 5,* 211–224. "Discussion." *SPEE, 5,* 224–246.

Merriman, M. (1893). Training of students in technical literary work. *Proceedings of the Society for the Promotion of Engineering Education, 1,* 259–264.

Miller, C.R. (1979). A humanistic rationale for technical writing. *College English, 40,* 610–617.

Miller, R.D. (1910). The teaching of English. *The Nation, 90,* 208.

Miller, W.J. (1975). What can the technical writer of the past teach the technical writer of today? In D.H. Cunningham & H.A. Estrin (Eds.), *The Teaching of Technical Writing*. (pp. 198–216). Urbana: NCTE.

Miller, W.J., & Saidla, L.E.A. (Eds.). (1953). *Engineers as writers: Growth of a literature*. New York: D. Van Nostrand.

Moran, M. (1985). The history of technical and scientific writing. In M.G. Moran & D. Journet (Eds.), *Research in technical communication: A bibliographical sourcebook*. (pp. 25–38). Westport, CT: Greenwood.

Moran, M.G. (1993). The road not taken: Frank Aydelotte and the thought approach to engineering writing. *Technical Communication Quarterly, 1*, 161–175.

Naether, C. (1935). Teaching the course in technical writing. *Proceedings of the Society for the Promotion of Engineering Education, 43*, 647–652.

Nelson, J.R. (1922). Conference of teachers of English. *Proceedings of the Society for the Promotion of Engineering Education, 30*, 255–282.

Nelson, J.R. (1940). *Writing the technical report*. New York: McGraw-Hill.

Newman, J.H. (1913). *Riverside essays: University subjects*. Cambridge: Riverside Press.

Newman, J.H. (1948). *The uses of knowledge*. New York: Appleton Century Crofts.

Noble, D.F. (1977). *America by design: Science, technology, and the rise of corporate capitalism*. New York: Knopf.

Oliver, L.M. (1940). *Technical exposition*. New York: McGraw-Hill.

Park, C.W. (1925). Some details of cooperation in the teaching of English to engineering students. *Proceedings of the Society for the Promotion of Engineering Education, 33*, 396–402.

Parks, C.W. (1919). Report of Committee No. 12, English. *Proceedings of the Society for the Promotion of Engineering Education, 27*, 318–326.

Parks, C.W. (1916). Report of the Committee on English. *Proceedings of the Society for the Promotion of Engineering Education, 24*, 177–182. Discussion. *SPEE, 24*, 183–193.

Perry, B. (1913). *The American mind and American idealism*. Cambridge, MA: Houghton Mifflin.

Pitman, J.H. (1948–49). A meeting of minds: The need for personal integration in the English and Humanities faculty. *Proceedings of the American Society for Engineering Education, 56*, 439–444.

Pitman, J.H. (1948–49). Selection of the faculty for the English department of an engineering college. *Proceedings of the American Society for Engineering Education, 56*, 318–322.

"Practical" Education. [Editorial]. (1912). *The Nation, 26*, 278–279.

Raymond, F. N. (1911). The preparation of written papers in schools of engineering. *Proceedings of the Society for the Promotion of Engineering Education, 19*, 48–54. Discussion. *SPEE, 19*, 54–90.

Rayner, W. H. (1922). The cultural element in engineering education. *Proceedings of the Society for the Promotion of Engineering Education, 30*, 153–161.

Rees, R.I. (1932). The young engineer and his English. *Proceedings of the Society for the Promotion of Engineering Education, 40*, 479–490.

Report of Committee on English. (1917). *Proceedings of the Society for the Promotion of Engineering Education, 25*, 210–225.

Report of Committee No. 12, English. (1922). *Proceedings of the Society for the Promotion of Engineering Education, 30*, 198–201.

Report of the Committee on English. (1927). *Proceedings of the Society for the Promotion of*

Engineering Education, 35, 109–111.

Report of the Conference on English. (1932). *Proceedings of the Society for the Promotion of Engineering Education, 40,* 782–797.

Report of the Conference on English. (1934). *Proceedings of the Society for the Promotion of Engineering Education, 42,* 244–245.

Report of the Conference on English. (1940). *Proceedings of the Society for the Promotion of Engineering Education, 48,* 302–304.

Report of the Conference on English. (1941). *Proceedings of the Society for the Promotion of Engineering Education, 49,* 276–280.

Report of the Conference on English. (1944). *Proceedings of the Society for the Promotion of Engineering Education, 52,* 94–95.

Report of the Investigation of Engineering Education, 1923–1929. (Wickenden Report, Vols. I and II). (1930). *Society for the promotion of engineering education.* Pittsburgh: University of Pittsburgh.

Reynolds, J.F. (1992). Classical rhetoric and the teaching of technical writing. *Technical Communication Quarterly, 1,* 63–76.

Reynolds, T.S., & Seely, B.E. (1993). Striving for balance: A hundred years of the American society for engineering education. *Journal of Engineering Education, 82,* 136–151.

Reynolds, T.S. (1992). The education of engineers in America before the Morrill Act of 1862. *History of Education Quarterly, 32,* 459–482.

Rickard, T.A. (1931). *Technical Writing* (3rd ed.). New York: Wiley.

Rivers, W.E. (1994). Studies in the history of business and technical writing: A bibliographical essay. *Journal of Business and Technical Communication, 8,* 6–57.

Robinson, A.T. (1909). The teaching of English in a scientific school. *Science, 30,* 657–664.

Rosner, M. (1983). Style and audience in technical writing: Advice from the early texts. *The Technical Writing Teacher, 11,* 38–45.

Russell, D.R. (1991). *Writing in the academic disciplines: 1870–1990: A curricular history.* Carbondale: Southern Illinois University Press.

Russell, D.R. (1993). The ethics of teaching ethics in professional communication: The case of engineering publicity at MIT in the 1920s. *Journal of Business and Technical Communication, 7,* 84–111.

Rutter, R. (1991). History, rhetoric, and humanism: Toward a more comprehensive definition of technical communication. *Journal of Technical Writing and Communication, 21,* 133–153.

Schelling, F.E. (1911). Discussion. *Society for the Promotion of Engineering Education, 19,* 54–90.

Schmelzer, R.W. (1977). The first textbook on technical writing. *Journal of Technical Writing and Communication, 7,* 51–54.

Seaver, H.L. (1910). Discussion. *Society for the Promotion of Engineering Education, 18,* 353–355.

Sinclair, B. (1986). Inventing a genteel tradition: MIT crosses the river. In B. Sinclair (Ed.), *New Perspectives on Technology and American Culture.* (APS Library Publication No. 12, pp. 1–18). Philadelphia: APS.

Souther, J. (1989). Teaching technical writing: A retrospective appraisal. In B.E. Fearing & W.K. Sparrow (Eds.), *Technical Writing.* (pp. 2–13). New York: MLA.

Smith, A.B. (1969). Historical development of concern for business English instruction. *Journal of Business Communication, 6*(3), 33–44.

Starbuck, A. (1924a). "Selling" English at Iowa State College. *Proceedings of the Society for the Promotion of Engineering Education, 32,* 334–341.

Starbuck, A. (1924b). The "Ames Narratives" and the "English Booth." *Proceedings of the Society for the Promotion of Engineering Education, 32,* 480–486.

Stetson, G.A. (1932). The art of technical writing. *Proceedings of the Society for the Promotion of Engineering Education, 40,* 491–498.

Stoughton, B., & Roe, M. (1924). Education in English for engineering students. *Proceedings of the Society for the Promotion of Engineering Education, 32,* 140–147.

Study of engineering education. (1923). *Science, 58,* 417.

Swain, P.W. (1944). English, the no. 1 vocational study for engineers. (Address given on June 23, 1944. Referenced in Report of the Conference on English.) *Proceedings of the Society for the Promotion of Engineering Education, 52,* 94–95.

Symposium on training of engineering teachers. (1921). *Proceedings of the Society for the Promotion of Engineering Education, 29,* 110–117.

Sypherd, W.O., & Brown, S. (1933). *The engineer's manual of English.* Chicago: Scott.

Sypherd, W.O. (1939). Thirty years of teaching English to engineers. *English Notes. Proceedings of the Society for the Promotion of Engineering Education, 47,* 161–165.

Telleen. (1908). Discussion. *Society for the Promotion of Engineering Education, 16,* 95–96.

Telleen, J.M. (1908). The courses in English in our technical schools. *Proceedings of the Society for the Promotion of Engineering Education, 16,* 61–73.

Thompson, K.O. (1945). English and the humanistic-social division. *Proceedings of the Society for the Promotion of Engineering Education, 53,* 163–169.

Thurber, S. (1893). The "Harvard Report" on the teaching of English. *Educational Review, 13,* 381–384.

Thurston, R.H. (1898). On the organization of engineering courses, and on entrance requirements for professional schools. *Proceedings of the Society for the Promotion of Engineering Education, 6,* 103–185.

Trelease, S.F., & Yule, E.S. (1937). *Preparation of scientific and technical papers* (3rd ed.). Baltimore: Williams and Wilkins.

Veblen, T. (1918). *The higher learning in America.* New York: Huebsch.

Veysey, L.R. (1965). *The emergence of the American university.* Chicago: University of Chicago Press.

Weeks, F. (1985). The teaching of business writing at the collegiate level, 1900–1920. In G.H. Douglas & H.W. Hildebrandt (Eds.), *Studies in the history of business writing.* (pp. 201–215). Urbana: Association for Business Communication.

Wickenden, W.E. (1927). The engineer's valuation of English. *English Journal, 16,* 446–453.

Wright, L.B. (1926). English and technical coordination at the University of North Carolina. *Proceedings of the Society for the Promotion of Engineering Education, 34,* 262–264.

Zappen, J.P. (1987). Historical studies in the rhetoric of science and technology. *The Technical Writing Teacher, 14,* 285–298.

Author Index

Subject Index

DATE DUE

Demco, Inc. 38-293